Bacterial Cellulose

This reference book provides updated information on the production and industrial significance of bacterial cellulose. Bacterial cellulose is a natural fiber produced by certain microbes, mainly bacteria which belong to the Acetobacter genera. The book discusses its applications in different industrial sectors, such as food, pharmaceutical, energy, and wastewater treatment. It covers the production of cellulose from conventional and renewable feedstock and includes topics such as downstream processing, characterization, and chemical modification of bacterial cellulose.

FEATURES

- Addresses the challenges of the production technologies of bacterial cellulose up to pilot scale
- Discusses cost-effective green processes using agri-processing residues and medium formulation
- Includes efficient preparation of nanocomposites using in vitro and in vivo methods
- Provides the latest applications of bacterial cellulose in the food and pharmaceuticals fields
- Reviews the production of bacterial cellulose from conventional feedstock such as sugars and starches

This book is designed for industry experts and researchers of applied microbiology, bioprocesses, and industrial microbiology.

Bacterial Cellulose

Production, Scale-up and Applications

Edited by

Vinod Kumar

Fermentation and Microbial Biotechnology Division
CSIR—Indian Institute of Integrative Medicine, Jammu, India

Saurabh Saran

Fermentation and Microbial Biotechnology Division
CSIR—Indian Institute of Integrative Medicine, Jammu, India

Ashok Pandey

Centre for Innovation and Translational Research
CSIR—Indian Institute of Toxicology Research, Lucknow, India

Carlos Ricardo Soccol

Department of Bioprocesses Engineering and Biotechnology
Federal University of Parana, Curitiba, Brazil

CRC Press
Taylor & Francis Group
Boca Raton London New York

CRC Press is an imprint of the
Taylor & Francis Group, an **informa** business

Designed cover image: Vinod Kumar

First edition published 2024
by CRC Press
2385 NW Executive Center Drive, Suite 320, Boca Raton FL 33431

and by CRC Press
4 Park Square, Milton Park, Abingdon, Oxon, OX14 4RN

CRC Press is an imprint of Taylor & Francis Group, LLC

ISBN: 978-1-032-39822-8 (hbk)
ISBN: 978-1-032-40939-9 (pbk)
ISBN: 978-1-003-35543-4 (ebk)

DOI: 10.1201/9781003355434

Typeset in Times
by Apex CoVantage, LLC

Contents

Preface

The discovery of cellulose by Payen in 1838 led to determining that it is a linear organic polysaccharide with hydrophilic behavior that is insoluble in water. Plants are the largest sources of cellulose, including from cotton fiber (which is about 90% cellulose), dried hemp (which contains around 40%–50% cellulose), and wood (which contains about 60% cellulose). Cellulose is well known for its commercial applications in the manufacturing of pulp, paper, textiles, and cellulose nanofibrous mats. Besides plants, cellulose can be found in various microbes, including fungi, bacteria, and algae. In 1886, Brown reported the first study of cellulose produced from bacteria, especially from *Acetobacter xylinum*. Recent research has shown that both Gram-positive bacteria such as *Sarcina ventriculi* and Gram-negative bacteria such as *Acetobacter* sp., *Azotobacter* sp., *Rhizobium* sp., *Agrobacterium* sp., *Pseudomonas* sp., *Salmonella* sp., and *Alcaligenes* sp. are capable of producing cellulose. The composition, structure, characteristics, and uses of the celluloses generated by various bacteria vary. *A. xylinum, A. hansenii*, and *A. pasteurianus* are the three bacteria that produce bacterial cellulose (BC) the most effectively. Due to its high productivity, *A. xylinum*, also known as *Gluconacetobacter xylinum*, has been utilized to produce BC economically.

Numerous sectors have developed innovative and alluring waste management strategies due to economic, social, and environmental factors. Integrated platforms such as sustainable bioengineering enable the material to be converted into useful, less hazardous final products. The use of BC as a significant biomaterial for numerous applications has a long history. However, because of the high fermentation cost, limited productivity, and expensive culture media, the potential for industrialization and commercialization of BC on a wide scale still faces difficulties. The latest developments being explored include increasing specific BC features by altering the growth conditions or using chemicals.

This book aims to provide basic and applied information related to BC, presenting the latest research and technological developments in a systematic manner. The book comprises 12 chapters. Chapter 1 presents some basic information about bacterial cellulose. The production of bacterial cellulose from conventional feedstock is covered in Chapter 2. Chapter 3 discusses the utilization of renewable feedstock for bacterial cellulose production. Chapter 4 presents the strategies for scaling up bacterial cellulose production. Chapter 5 discusses the recovery of bacterial cellulose from fermentation broth. Chapter 6 deals with the characterization and storage of bacterial cellulose. In Chapter 7, the chemical modification of bacterial cellulose for industrial applications is discussed. Chapter 8 deals with the in situ preparation of bacterial cellulose. Chapter 9 covers the preparation of bacterial cellulose nanocomposites and their characterization. Chapter 10 examines the application of bacterial cellulose matrices in various fields. Chapter 11 discusses bacterial cellulose applications in tissue regeneration and engineering, wound dressing, skin grafting, biosensors and diagnostics, drug delivery, water purification, and air filtration, especially in the nano form. Chapter 12 presents a discussion of the increasing fascination with bacterial cellulose–based composites, their versatility as sustainable biopolymers to strengthen the standards and requirements of the present age of biomedical materials. The contents have been written so that they can be understood by nonspecialists as well as those in related research fields. Each chapter conveys the latest trend in bacterial cellulose science and technology.

We sincerely appreciate the contributions of the scientific experts who shared their knowledge of bacterial cellulose. We strongly believe that this book provides enriched scientific information that will be useful for researchers, students, academicians, and industry experts. We are grateful to the reviewers for their sincere efforts and acknowledge them for their contribution in the critical review of the chapters. We sincerely acknowledge the Biotech Research Society, India (BRSI) for providing us the opportunity to prepare this book and guiding us during the editing and publication process to bring this book into its final form. We thank the team of Taylor & Francis Group, specifically

Dr Gagandeep Singh, Senior Publisher; Dr Madhurima Kahali, former Editor II (Life Sciences); Ms Neha Bhatt, Editorial Assistant; and the entire team at CRC Press—Taylor & Francis Group for their consistent support during the publication process of this book.

Vinod Kumar
Saurabh Saran
Ashok Pandey
Carlos Ricardo Soccol

Editors

Vinod Kumar, PhD, is a Scientist at the Fermentation and Microbial Biotechnology Division, CSIR—Indian Institute of Integrative Medicine, Jammu, India. He earned a PhD at the University of Delhi. Dr. Kumar was awarded the EPFL Visiting Fellowship in 2016. His current research interests are the production of bacterial cellulose and the preparation of a composite of bacterial cellulose, xylitol, and industrially important enzymes from unutilized bioresources and processing residues. He has published 42 papers and has four patents.

Saurabh Saran, PhD, is a Principal Scientist at the CSIR—Indian Institute of Integrative Medicine, Jammu, India. He earned a PhD at Delhi University, India. He works on strain engineering, strain development, biotransformation using CYP 450, upstream and downstream processing of different fermentation processes, and applications of these molecules in various industries. He has worked as a Research Scientist at Reliance Industries, Mumbai, and as a Research Professor in the Republic of Korea and Finland. He has seven patents, two books, and 45 publications to his credit.

Ashok Pandey, PhD, is currently a Distinguished Scientist at the Centre for Innovation and Translational Research, CSIR—Indian Institute of Toxicology Research, Lucknow, India, and Executive Director (Honorary) at the Centre for Energy and Environmental Sustainability, India. His major research and technological development interests include industrial and environmental biotechnology, energy biosciences, focusing on biomass to biofuels and chemicals, waste to wealth and energy, and industrial enzymes. He has 16 patents, 120 books and > 1000 papers and book chapters.

Carlos Ricardo Soccol, PhD, is the Research Group Leader of the Department of Bioprocesses Engineering and Biotechnology (DEBB) at the Federal University of Parana, Curitiba, Brazil. Professor Soccol is an expert in bioprocesses, biological engineering, industrial biotechnology, green fuels technology, applied microbiology, and fermentation technology. He has written over 1301 publications, including 101 patents, 26 books, 165 book chapters, and 425 original papers. Dr. Soccol has supervised 83 PhD students. He was a Visiting Professor at the École Polytechnique Fédérale de Lausanne, Switzerland, University Lille, France, and University Blaise Pascal, Clermont Ferrand, France. He is a technical and scientific consultant for several companies and agencies.

Contributors

Amitabh Aharwar
Government College, Harrai
Chhindwara, Madhya Pradesh, India

Vivek Kumar Ahluwalia
Center of Innovative and Applied Bioprocessing (CIAB)
Mohali, Punjab, India

Vasudha Bansal
Department of Foods and Nutrition
Government Home Science College (Affiliated with Panjab University, Chandigarh)
Chandigarh, Punjab and Haryana, India

Sandesh Behera
Department of Biotechnology
Maharaja Sriram Chandra Bhanja Deo University
Baripada, Odisha, India

Archana Bhaskar
Rajiv Gandhi Arts and Commerce College
Mungeli, Chhattisgarh, India

Parameswaran Binod
CSIR—National Institute for Interdisciplinary Science and Technology
Trivandrum, Kerala, India

Yogita V. Chavan
MIT School of Food Technology
MIT Art, Design and Technology University
Pune, Maharashtra, India

Alexander da Silva Vale
Department of Bioprocess Engineering and Biotechnology
Federal University of Paraná
Curitiba, Paraná, Brazil

Júlio Cesar de Carvalho
Department of Bioprocess Engineering and Biotechnology
Federal University of Paraná
Curitiba, Paraná, Brazil

Rafaela de Oliveira Penha
Department of Bioprocess Engineering and Biotechnology
Federal University of Paraná
Curitiba, Paraná, Brazil

Aida Fadakar Sarkandi
Department of Textile Engineering
Functional Fibrous Structures and Environmental Enhancement (FFSEE)
Amirkabir University of Technology
Tehran, Iran

Leonardo Wedderhoff Herrmann
Department of Bioprocess Engineering and Biotechnology
Federal University of Paraná
Curitiba, Paraná, Brazil

Susan Grace Karp
Department of Bioprocess Engineering and Biotechnology
Federal University of Paraná
Curitiba, Paraná, Brazil

Farheen Khan
Regulatory Toxicology Group
Centre for Innovation and Translational Research
Council of Scientific and Industrial Research (CSIR)—Indian Institute of Toxicology Research (IITR)
Lucknow, Uttar Pradesh, India

Apostolis Koutinas
Department of Food Science and Human Nutrition
Agricultural University of Athens
Athens, Greece

Manoj Kumar
Fermentation and Microbial Biotechnology Division
Council of Scientific and Industrial Research (CSIR)—Indian Institute of Integrative Medicine (IIIM)
Jammu, Jammu and Kashmir, India
and
Academy of Scientific and Innovative Research (AcSIR)
Ghaziabad, Uttar Pradesh, India

Vikas Kumar
Department of Food Science and Technology
Punjab Agricultural University
Ludhiana, Punjab, India

Vinod Kumar
Fermentation and Microbial Biotechnology Division
CSIR—Indian Institute of Integrative Medicine
Jammu, Jammu & Kashmir, India
and
Academy of Scientific and Innovative Research (AcSIR)
Ghaziabad, Uttar Pradesh, India

Luiz Alberto Junior Letti
Department of Bioprocess Engineering and Biotechnology
Federal University of Paraná
Curitiba, Paraná, Brazil

Walter José Martínez-Burgos
Department of Bioprocess Engineering and Biotechnology
Federal University of Paraná
Curitiba, Paraná, Brazil

Majid Montazer
Department of Textile Engineering
Functional Fibrous Structures and Environmental Enhancement (FFSEE)
and
Amirkabir Nanotechnology Research Institute (ANTRI)
Amirkabir University of Technology
Tehran, Iran

Ashok Pandey
Centre for Innovation and Translational Research
CSIR—Indian Institute of Toxicology Research, Lucknow, India

Parmjit S. Panesar
Food Biotechnology Research Laboratory
Department of Food Engineering and Technology
Sant Longowal Institute of Engineering and Technology
Longowal, Punjab, India

Dharmendra Kumar Parihar
Department of Biotechnology
Guru Ghasidas Vishwavidyalaya
Bilaspur, Chhattisgarh, India

Anil Kumar Patel
Institute of Aquatic Science and Technology
National Kaohsiung University of Science and Technology
Kaohsiung, Taiwan

Olga Psaki
Department of Food Science and Human Nutrition
Agricultural University of Athens
Athens, Greece

Anu Radha
Fermentation and Microbial Biotechnology Division
Council of Scientific and Industrial Research (CSIR)—Indian Institute of Integrative Medicine (IIIM)
Jammu, Jammu and Kashmir, India

Amit Kumar Rai
Microbial Resources
Institute of Bioresources and Sustainable Development
Gangtok, Sikkim, India

Cristine Rodrigues
Department of Bioprocess Engineering and Biotechnology
Federal University of Paraná
Curitiba, Paraná, Brazil

Mirva Sarafidou
Department of Food Science and Human Nutrition
Agricultural University of Athens
Athens, Greece

Saurabh Saran
Fermentation and Microbial Biotechnology Division
Council of Scientific and Industrial Research (CSIR)—Indian Institute of Integrative Medicine (IIIM)
Jammu, Jammu and Kashmir, India
and
Academy of Scientific and Innovative Research (AcSIR)
Ghaziabad, Uttar Pradesh, India

Thamarys Scapini
Department of Bioprocess Engineering and Biotechnology
Federal University of Paraná
Curitiba, Paraná, Brazil

Nitya Sharma
Centre for Rural Development and Technology
Indian Institute of Technology Delhi
New Delhi, India

Vinod Pravin Sharma
Regulatory Toxicology Group
Centre for Innovation and Translational Research
Council of Scientific and Industrial Research (CSIR)—Indian Institute of Toxicology Research (IITR)
Lucknow, Uttar Pradesh, India

Raveendran Sindhu
Department of Food Technology
Thangal Kunju Musaliar Institute of Technology
Kollam, Kerala, India

Reeta Rani Singhania
Department of Marine Environmental Engineering
National Kaohsiung University of Science and Technology
Kaohsiung, Taiwan

Anupreet Kaur Sobti
Department of Foods and Nutrition
Government Home Science College (Affiliated with Panjab University, Chandigarh)
Chandigarh, Punjab and Haryana, India

Carlos Ricardo Soccol
Department of Bioprocess Engineering and Biotechnology
Federal University of Paraná
Curitiba, Paraná, Brazil

Vanete Thomaz-Soccol
Department of Bioprocess Engineering and Biotechnology
Federal University of Paraná
Curitiba, Paraná, Brazil

Vartika Srivastava
Regulatory Toxicology Group
Centre for Innovation and Translational Research
Council of Scientific and Industrial Research (CSIR)—Indian Institute of Toxicology Research (IITR)
Lucknow, Uttar Pradesh, India

Hrudayanath Thatoi
Department of Biotechnology
Maharaja Sriram Chandra Bhanja Deo University
Baripada, Odisha, India

Sunita Varjani
School of Energy and Environment
City University of Hong Kong
Kowloon, Hong Kong
and
Sustainability Cluster
School of Engineering
University of Petroleum and Energy Studies
Dehradun, Uttarakhand, India

Bhoomika Varshney
Regulatory Toxicology Group
Centre for Innovation and Translational Research
Council of Scientific and Industrial Research (CSIR)—Indian Institute of Toxicology Research (IITR)
Lucknow, Uttar Pradesh, India

Adenise Lorenci Woiciechowski
Department of Bioprocess Engineering and Biotechnology
Federal University of Paraná
Curitiba, Paraná, Brazil

1 Bacterial Cellulose—An Introduction

Carlos Ricardo Soccol, Luiz Alberto Junior Letti, Susan Grace Karp, Cristine Rodrigues, Vanete Thomaz-Soccol, Júlio Cesar de Carvalho, Leonardo Wedderhoff Herrmann, Alexander da Silva Vale, and Adenise Lorenci Woiciechowski
Federal University of Paraná
Curitiba, Parana, Brazil

1.1 INTRODUCTION

Cellulose is the most abundant biopolymer in the world, with applications in several areas, including the paper industry, construction, pharmaceuticals, medicine, among others. It is chemically constituted of linear chains of poly-β-(1,4)-D-glucose joined via glycosidic linkages, the repeating unit of this polymer being the cellobiose molecule, formed by the union of two glucose molecules. Cellulose can be obtained from plants, algae, and microorganisms and can be synthesized enzymatically (Manan et al. 2022).

The widely commercialized source of this biopolymer is the cell wall of plants; however, plant cellulose is present with other biopolymers, such as lignin and hemicelluloses, which makes the use of harsh chemicals necessary for its extraction. Furthermore, in the vegetal cellulose fiber, the molecules of cellulose are arranged in a very stable structure, stabilized by molecular interaction with intra- and intermolecular hydrogen bonds, producing a very recalcitrant structure. The geometry of bacterial cellulose (BC) molecules is determined by the network of intramolecular and intermolecular hydrogen bonds, hydrophobic and van der Waals interactions, forming parallel chains called cellulose microfibrils (Koizumi et al. 2008). With the possibility of different modifications (chemical or enzymatic), BC becomes quite versatile, being obtained as micro- or nanofibrils with different degrees of crystallinity (Cacicedo et al. 2016).

In this sense, BC has advantages in being produced extracellularly and released in a purer form in the medium, reducing the number of steps in the purification process. In addition, although it has the same chemical composition as plant cellulose, BC has different properties, such as crystallinity, increased water absorption capacity, resistance to traction, and a high degree of polymerization (Singhania et al. 2022).

1.2 PRODUCER MICROORGANISMS

Bacterial cellulose was first noticed and reported by Brown (1886) while studying chemical reactions in the static acetic fermentation process with the bacterium *Acetobacter xylinum*. Brown (1886) noticed the formation of a gelatinous film on the surface of the fermentation medium. Its existence, however, had been observed before, when it was called the "mother vinegar."

Several bacterial genera are capable of producing cellulose (usually to protect themselves against a chemically aggressive environment and ultraviolet radiation), such as *Acetobacter, Glucanoacetobacter, Azotobacter, Pseudomonas, Salmonella, Sarcina, Agrobacterium, Aerobacter, Achromobacter, Enterobacter, Coronobacter, Rhizobium, Clostridium, Alcaligenes*, among others. However, the genus *Komagataeibacter* (formerly classified as *Acetobacter* or *Glucanoacetobacter*)

DOI: 10.1201/9781003355434-1

is the most used due to its high yield and productivity, especially *Komagataeibacter xylinus*. Cellulose biosynthesis is well established in such strains, comprising a chain of complex reactions involving four main enzymes. Still, recent studies have highlighted the importance of detailed knowledge of metabolic pathways to optimize cellulose production, lower process costs, and unveil new operons to allow posttranscription reactions and the generation of biobased materials with differentiated properties for the most diverse industrial applications (Singhania et al. 2022; Manan et al. 2022).

1.3 PRODUCTION OF BACTERIAL CELLULOSE

Bacterial cellulose is often produced via aerobic fermentation as a surface pellicle in static cultivation. However, agitated cultures were also reported, yielding spherical or irregularly shaped particles with lower crystallinity, lower degree of polymerization and lower mechanical strength, as compared to the film produced in the static condition. In both static and agitated cultures, one of the greatest challenges in large-scale processes is the oxygen supply (Singhania et al. 2022).

Strategies to allow the economic production of BC at commercial scale are dependent on the production mode, and usually involve the use of cheaper carbon and nitrogen sources, the design of appropriate bioreactors, and the optimization of operation modes (Singhania et al. 2022).

1.3.1 Media Sources

The Hestrin-Schramm (HS) medium, composed of glucose, yeast extract, peptone, disodium hydrogen phosphate, and citric acid, is the most popular for the production of BC. However, several carbon and nitrogen sources have been employed in fermentation processes for BC production, including glucose, sucrose, fructose, maltose, crude glycerol, syrups, biomass hydrolysates, corn steep liquor, whey, beef extract, peptone, yeast extract, and ammonium sulfate, with product yields ranging from 0.1 to 18 g/L (Singhania et al. 2022).

Sugar hydrolysates obtained from plant biomass, such as grass straw, grass husk, wheat husk, and corn cob were evaluated as substitutes for commercial carbon sources in HS medium in the production of bacterial nanocellulose, as a strategy of valorization of lignocellulosic wastes, and the product yields varied between 0.9 and 1.5 g/L (Sadalage and Pawar 2021). Recently, detoxified sugarcane bagasse hydrolysate was used as a medium for BC production in static cultivation. The sugar hydrolysate, prepared by sulfuric acid hydrolysis, was detoxified by atmospheric cold plasma treatment, and yielded 1.68 g/L of BC, while the untreated hydrolysate yielded 1.88 g/L. This indicated that the detoxification process may have removed important nutrients from the culture broth. The highest production, above 2 g/L, was obtained in modified HS medium, an HS-based medium with its carbon source adjusted to the same level as that of the plasma-untreated medium (Lin et al. 2022).

1.3.2 Production Strategies

In static cultivation, the most targeted parameters to improve BC production are the carbon source, as well as the operation mode, air supply, and surface area. A study developed by Basu, Vadanan, and Lim (2019), that aimed at finding significant parameters for a scalable bacterial cellulose production platform under static conditions, demonstrated that the carbon source and the optimal reactor design were the most important factors for obtaining improved yields. The projected yields of BC were as high as 40 g/L for *Gluconacetobacter hansenii* 53582 grown on sucrose as the carbon source.

The semicontinuous process was recommended for industrial production in static condition (Çakar et al. 2014), to avoid the depletion of the carbon source and the accumulation of organic acids that reduce the pH. Ethanol supplementation of the medium also showed a positive effect on static BC production (Singhania et al. 2022). Concerning oxygen supply, one strategy reported in the literature to improve oxygen mass transfer to the fermentation medium was the use of nano-gas bubbled water (Patel et al. 2021).

The production of BC in static cultivation under semi-anaerobic conditions was accomplished when hydrocolloids such as xanthan gum and konjac glucomannan were added to the cultivation medium, triggering cellulose metabolism in *Acetobacter xylinum* via the conventional aerobic pathway (Nguyen et al. 2022). Another strategy applied to increase BC yield in static cultivation was the application of external electromagnetic fields, specifically a rotating magnetic field, to stimulate bacterial growth and improve oxygen mass transfer to the culture medium. Increased metabolic activity, higher uptake of substrates (fructose, ethanol, and citric acid), higher biomass production, and an increase of more than 30% in BC production were observed under rotating magnetic field sequential exposure (Konopacki et al. 2022).

For industrial-scale production, however, stirred or agitated fermentation is still recommended. Here, it is important to adjust the fermentation parameters to regulate the shape and the properties of the synthesized cellulose particles. These parameters include reactor design, carbon source concentration, inoculum size, and optical density (Singhania et al. 2022).

The production of BC in agitated cultures is characterized by an increase in medium viscosity during fermentation, which brings challenges such as oxygen transfer and agitation power supply (Shoda and Sugano 2005). One alternative to overcome these obstacles is the use of airlift reactors, which are more energy efficient and promote less shear stress as compared to stirred tank reactors. A rotating-disc bioreactor, which maintains the bacteria in a rotating disk immersed in the medium, exposed to the atmosphere, and allows the formation of biofilms, is another option to produce BC at large scale. Additionally, the limitation in oxygen mass transfer can be minimized by the use of pure oxygen or oxygen-enriched air instead of air (Singhania et al. 2022).

1.4 APPLICATIONS OF BACTERIAL CELLULOSE

In the food industry, the classic product obtained from BC is nata de coco, and several carbon sources can be used, either conventional or renewable feedstock. The most successful biomedical application is the use of BC films for wound dressing and as a substitute for skin (artificial skin), after a suitable downstream procedure, due to its mechanical characteristics and its ability to provide an efficient barrier against microbial infections, while allowing air transport through the membrane. Other advanced applications can be cited in other areas, such as the paper industry (phosphorylated BC can be employed for the production of paper with high strength and low smoke generation), detergent industry (BC as surfactant and thickener for liquid laundry detergents), leather industry (BC used for production of resilient leather-like materials), biomedicine (BC scaffolds with micropores have been used for bone tissue engineering), cosmetic industry (the insertion of biocompounds from plant extracts in BC allow the production of face masks), among several other potential uses (including as a membrane for filtration technology, drug delivery, energy cells and fuel cell membranes, nanocellulose for food packaging, bioactive paper used in clinical diagnosis, texture enhancer, antioxidant and antibacterial agent, immobilization of enzymes and metal ions, blood tubes, hearth valves) (Basta and El-Saied 2009; García and Prieto 2019; Wang, Tavakoli, and Tang 2019; Andriani, Apriyana, and Karina 2020; Fernandes et al. 2020). These and other applications are detailed in the final chapters of this book.

1.5 DEMAND OF BACTERIAL CELLULOSE

The demand for BC derivate products has grown rapidly in recent years and estimates for its global market value in 2026 are around US $500 million. However, the commercial production of bacterial cellulose has not yet reached its full potential because of some major drawbacks, including the low yield and the high production costs (Singhania et al. 2022). Nevertheless, the interest in BC-based technologies is growing, as evidenced by the number of patents filed.

Patents related to "bacterial cellulose" totaled 2,763 documents in the Derwent Innovations Index Database; almost 50% of them published in the last five years. Of these, 1,212 documents, or 44%, were filed by universities and research institutes, the top assignees being the Chinese Donghua University, the Nanjing University of Science and Technology, and the Tianjin University of Science and Technology. Among companies, the Japanese Biopolymer Research KK was the top assignee, with 29 filed patent documents, most of them related to methods for the preparation of bacterial cellulose through fermentation processes; novel microorganisms; new bioreactor models; methods for treating bacterial cellulose solutions and suspensions; and bacterial cellulose-based products, including paper, food ingredients, and cosmetics. Other frequent assignees were Hainan Guangyu Biotechnology Co. Ltd. (China), Hainan Yeguo Foods Co. Ltd. (China), China Tobacco Henan Industrial Co. Ltd. (China), Weyerhaeuser Company (US), Shandong Nameide Biotechnology Co. Ltd. (China), and Procter & Gamble Co. (US).

The most recent patent document, filed by the Zhende Medical Co. Ltd. (China) and published in June 2022, presented a utility model of a BC dressing, provided with a moisture-preserving film and moisture-indicating layer, containing cobalt chloride as a water indicator. The utility model allows prolonged use of the dressing and indicates to the user when it is time to change the dressing, according to a color change (Lu and Wang 2022).

Another document filed by the Fujian Hengan Group Co. Ltd. (China) and published in March 2022 described a self-suspending, long-acting, green detergent comprising bacterial cellulose and enzymes (protease, amylase, cellulase, lipase, mannanase, and/or pectinase) (Yu and Zhuang 2022).

An alternative application was proposed in a patent document filed by the Ningbo Institute of Materials Technology and Engineering (China), published in March 2022, a composite film to be applied in a lithium-ion battery diaphragm useful in vehicle and energy storage industries. The composite film is comprised of BC and a composite powder containing sodium alginate, carboxymethyl cellulose, and/or polyacrylic acid mixed with an inorganic substance (diatomite, zeolite, kaolin, and/or montmorillonite). The composite film has good, long-term electrochemical performance, strong mechanical strength, good acid and alkali corrosion resistance, and good wetting property, among other advantages (Guo and He 2022).

1.6 CONCLUSIONS

Bacterial cellulose is an important microbial metabolite that, due to its particular way of being synthesized, produces cellulose in single molecules, free from other associated structures, unlike the cellulose found in vegetal cells. This particular product, due to the final structure form, is attracting much attention and much research is being developed by research groups all over the world to find new and more productive microorganisms, to define new production strategies with alternative substrates in order to increase the product quality and decrease the production costs. These advances are important because bacterial cellulose can be used in many important industrial and medical applications summarized in this chapter, which will be discussed in the following chapters.

1.7 REFERENCES

Andriani, Dian, Arina Yuthi Apriyana, and Myrtha Karina. 2020. "The Optimization of Bacterial Cellulose Production and Its Applications: A Review." *Cellulose* 27 (12). Springer: 6747–6766. doi:10.1007/S10570-020-03273-9.

Basta, A. H., and H. El-Saied. 2009. "Performance of Improved Bacterial Cellulose Application in the Production of Functional Paper." *Journal of Applied Microbiology* 107 (6). J Appl Microbiol: 2098–2107. doi:10.1111/J.1365-2672.2009.04467.X.

Basu, Anindya, Sundaravadanam Vishnu Vadanan, and Sierin Lim. 2019. "Rational Design of a Scalable Bioprocess Platform for Bacterial Cellulose Production." *Carbohydrate Polymers* 207 (March). Elsevier: 684–693. doi:10.1016/J.CARBPOL.2018.10.085.

Brown, Adrian J. 1886. "XLIII: On an Acetic Ferment Which Forms Cellulose." *Journal of the Chemical Society, Transactions* 49 (0). The Royal Society of Chemistry: 432–439. doi:10.1039/CT8864900432.

Cacicedo, Maximiliano L., M. Cristina Castro, Ioannis Servetas, Loulouda Bosnea, Konstantina Boura, Panagiota Tsafrakidou, Agapi Dima, Antonia Terpou, Athanasios Koutinas, and Guillermo R. Castro. 2016. "Progress in Bacterial Cellulose Matrices for Biotechnological Applications." *Bioresource Technology* 213 (August). Elsevier: 172–180. doi:10.1016/J.BIORTECH.2016.02.071.

Çakar, Fatih, Işilay Özer, A. Özhan Aytekin, and Fikrettin Şahin. 2014. "Improvement Production of Bacterial Cellulose by Semi-Continuous Process in Molasses Medium." *Carbohydrate Polymers* 106 (1). Elsevier: 7–13. doi:10.1016/J.CARBPOL.2014.01.103.

Fernandes, Isabela de Andrade Arruda, Alessandra Cristina Pedro, Valéria Rampazzo Ribeiro, Débora Gonçalves Bortolini, Mellany Sarah Cabral Ozaki, Giselle Maria Maciel, and Charles Windson Isidoro Haminiuk. 2020. "Bacterial Cellulose: From Production Optimization to New Applications." *International Journal of Biological Macromolecules* 164 (December). Elsevier: 2598–2611. doi:10.1016/J.IJBIOMAC.2020.07.255.

García, Concha, and María Auxiliadora Prieto. 2019. "Bacterial Cellulose as a Potential Bioleather Substitute for the Footwear Industry." *Microbial Biotechnology* 12 (4). John Wiley & Sons, Ltd: 582–585. doi:10.1111/1751-7915.13306.

Guo, Z., and H. He. 2022. Composite Film Useful for Battery Diaphragm of Lithium Ion Battery, Comprises Bacterial Cellulose and Composite Powder Comprising Organic Macromolecular Compound Comprising at Least One of Sodium Alginate, Carboxymethyl Cellulose and/or Polyacrylic Acid a. CN114256560-A, issued 2022.

Koizumi, S., Z. Yue, Y. Tomita, T. Kondo, H. Iwase, D. Yamaguchi, and T. Hashimoto. 2008. "Bacterium Organizes Hierarchical Amorphous Structure in Microbial Cellulose." *The European Physical Journal E* 26 (1). Springer: 137–142. doi:10.1140/EPJE/I2007-10259-3.

Konopacki, Maciej, Bartłomiej Grygorcewicz, Marian Kordas, Paula Ossowicz-Rupniewska, Anna Nowak, Magdalena Perużyńska, and Rafał Rakoczy. 2022. "Intensification of Bacterial Cellulose Production Process with Sequential Electromagnetic Field Exposure Aided by Dynamic Modelling." *Biochemical Engineering Journal* 182 (May). Elsevier: 108432. doi:10.1016/J.BEJ.2022.108432.

Lin, Shin Ping, Shyh Haur Huang, Yuwen Ting, Hsien Yi Hsu, and Kuan Chen Cheng. 2022. "Evaluation of Detoxified Sugarcane Bagasse Hydrolysate by Atmospheric Cold Plasma for Bacterial Cellulose Production." *International Journal of Biological Macromolecules* 204 (April). Elsevier: 136–143. doi:10.1016/J.IJBIOMAC.2022.01.186.

Lu, J., and J. Wang. 2022. Bacterial Cellulose Dressing Comprises Bacterial Cellulose Base Material Adjacent to Contact Surface and Non Contact Surface, Where Non-Contact Wound Surface is Provided with Water Indicating Layer, Which is Water Indicator. CN216652630-U, issued 2022.

Manan, Sehrish, Muhammad Wajid Ullah, Mazhar Ul-Islam, Zhijun Shi, Mario Gauthier, and Guang Yang. 2022. "Bacterial Cellulose: Molecular Regulation of Biosynthesis, Supramolecular Assembly, and Tailored Structural and Functional Properties." *Progress in Materials Science* 129 (August). Pergamon: 100972. doi:10.1016/J.PMATSCI.2022.100972.

Nguyen, Quoc Duy, Thi Van Linh Nguyen, Thi Thuy Dung Nguyen, and Nhu Ngoc Nguyen. 2022. "Effects of Different Hydrocolloids on the Production of Bacterial Cellulose by Acetobacter Xylinum Using Hestrin: Schramm Medium under Anaerobic Condition." *Bioresource Technology Reports* 17 (February). Elsevier: 100878. doi:10.1016/J.BITEB.2021.100878.

Patel, Anil Kumar, Reeta Rani Singhania, Chiu Wen Chen, Yi Sheng Tseng, Chia Hung Kuo, Chien Hui Wu, and Cheng Di Dong. 2021. "Advances in Micro-and Nano Bubbles Technology for Application in Biochemical Processes." *Environmental Technology & Innovation* 23 (August). Elsevier: 101729. doi:10.1016/J.ETI.2021.101729.

Sadalage, Priyadarshani S., and Kiran D. Pawar. 2021. "Production of Microcrystalline Cellulose and Bacterial Nanocellulose through Biological Valorization of Lignocellulosic Biomass Wastes." *Journal of Cleaner Production* 327 (December). Elsevier: 129462. doi:10.1016/J.JCLEPRO.2021.129462.

Shoda, Makoto, and Yasushi Sugano. 2005. "Recent Advances in Bacterial Cellulose Production." *Biotechnology and Bioprocess Engineering* 10 (1). Springer: 1–8. doi:10.1007/BF02931175.

Singhania, Reeta Rani, Anil Kumar Patel, Yi Sheng Tseng, Vinod Kumar, Chiu Wen Chen, Dibyajyoti Haldar, Jitendra Kumar Saini, and Cheng Di Dong. 2022. "Developments in Bioprocess for Bacterial Cellulose Production." *Bioresource Technology* 344 (January). Elsevier: 126343. doi:10.1016/J.BIORTECH.2021.126343.

Wang, Jing, Javad Tavakoli, and Youhong Tang. 2019. "Bacterial Cellulose Production, Properties and Applications with Different Culture Methods: A Review." *Carbohydrate Polymers* 219 (September). Elsevier: 63–76. doi:10.1016/J.CARBPOL.2019.05.008.

Yu, C., and Q. Zhuang. 2022. Self-Suspending, Long-Acting and Green Detergent Comprises Water, Sodium Fatty Alcohol Polyoxyethylene Ether Carboxylate, Sodium Alcohol Ether Carboxylate, Cocamidopropyl Betaine, Tetrasodium Glutamate Diacetate, Bacterial Cellulose, Biological Enzyme Gel. CN114250122-A, issued 2022.

2 Production of Bacterial Cellulose from Conventional Feedstock

Adenise Lorenci Woiciechowski, Rafaela de Oliveira Penha, Thamarys Scapini, Susan Grace Karp, Cristine Rodrigues, Júlio Cesar de Carvalho, Vanete Thomaz-Soccol, Walter José Martínez-Burgos, and Carlos Ricardo Soccol
Federal University of Paraná
Curitiba, Paraná, Brazil

2.1 INTRODUCTION

First reported more than a century ago by Brown (1886), bacterial cellulose (BC) was produced by *Bacterium xylinum* (also named *Acetobacter xylinus* and *Gluconacetobacter xylinus*) (Yamada 1983), and recently accommodated to the *Komagataeibacter* group, now denominated *Komagataeibacter xylinus* (Yamada et al. 2012) and described as a "jelly-like translucent mass on the surface of the culture fluid." After years of technological development and BC production, it is strongly suggested that the BC is an important biopolymer that fulfills different biological roles in the natural habitat of the producing microorganisms, providing mechanical, chemical, and physiological protection; it can act in the fixation, adhesion, and substrate colonization, and protecting the cells against ultraviolet radiation and abrupt changes of factors in the habitat (e.g., pH, toxic compounds, water content) (Ross et al. 1991).

BC is an exclusively extracellular bioproduct synthesized from monomeric carbon molecules, composed of β-1,4-linked D-glucose products, regulated by different metabolic pathways involving several individual enzymes, catalytic complexes, and regulatory proteins (Ross et al. 1991; Fernández et al. 2019). The technological interest in efficient BC production is mainly due to the chemical purity of the biocompound that is free of hemicellulose, lignin, and other compounds present in plant cellulose. In addition, it has high structural crystallinity, high water retention capacity, elasticity, tensile strength, and it is biocompatible, which expands the biotechnological potential of BC, being a multifunctional biomaterial applied mainly in the biomedical field, catalysis, conductive materials, and electrical devices (Fernández et al. 2019; Gorgieva and Trček 2019).

Various microorganisms have been successfully employed in BC production, such as *Acetobacter*, *Azotobacter*, *Rhizobium*, *Agrobacterium*, *Pseudomonas*, *Salmonella*, *Alcaligenes*, and *Sarcina ventriculi* (Wang et al. 2019). However, the production pathways are well described for strictly aerobic Gram-negative bacteria *Komagataeibacter xylinus* (relocated from the genera *Gluconacetobacter* and *Acetobacter*) which are the most widely employed in BC production and have been widely described in the literature.

In this sense, beyond establishing the producing microorganism, research has been conducted to optimize the culture medium (mainly carbon and nitrogen sources) and environmental factors crucial for cost-effective process performance. The carbon source used for BC production can affect different BC properties, such as degree of polymerization (DP), surface area, porosity, water retention capacities, molecular weight, crystallinity index, mechanical properties, and intrinsic viscosity (Gorgieva and Trček 2019). The culture medium directly affects BC production; it is estimated

DOI: 10.1201/9781003355434-2

that the cost of the medium corresponds to about 30% of the total cost of production, being one disadvantage of BC production on a large scale (Jozala et al. 2016; Kosseva et al. 2020). The most well-known synthetic medium for BC production is the Hestrin-Schramm (HS) medium which has glucose as its main carbon source. To overcome the challenge, alternative carbon sources to the synthetic medium are being explored, such as juice from sugarcane, sugar beets, fruits, and other feedstock conventionally applied for biotechnological fermentative processes.

Based on that, this chapter addresses the advances in BC production, including discussions on the utilization of synthetic culture media and their substitution by conventional feedstock.

2.2 CONVENTIONAL FEEDSTOCK: A STRATEGY FOR BACTERIAL CELLULOSE PRODUCTION

The advances in research in BC production have been relevant to expand the potential application of the bioproduct, and with those significant advances also viable industrial processes have been established, with satisfactory yield and biocompatible products for different purposes. In this sense, the development of a robust bioprocess for BC production must establish a synergy between a range of components that directly affect the yield of the system and the bioproduct properties, such as the environmental factors (pH, temperature, working volume, dissolved oxygen) and, mainly, the culture medium.

Traditionally, BC production is performed using a synthetic culture medium, mostly HS medium (Hestrin and Schramm 1954) composed of glucose (2% v v^{-1}) as a carbon source and other nutrients sources, such peptone (0.5% v v^{-1}), yeast extract (0.5% v v^{-1}), disodium phosphate (0.27% v v^{-1}), and citric acid (0.115% v v^{-1}), at pH 6.0. Other carbon sources such as fructose, mannitol, maltose, xylose, sucrose, galactose, and others have been tested to optimize the BC yield (see Table 2.1 in Section 2.3.1). Despite presenting attractive yields, and important process control for the BC application in some fields, such as pharmaceutical and medical fields, the use of synthetic media limits the expansion of the bioproduct to industrial scale in a range of other applications because the process costs become high and would result in biomaterials that are not economically profitable.

In this context, cheaper feedstock from carbon sources and other nutrients have been evaluated as a strategy to overcome the limitation and increase the commercial competitiveness of BC. Conventional feedstock are already widely used in the industry for their high content of sugars that are converted into various products, such as first-generation biofuels, functional foods, and organic acids and other compounds from sugarcane, corn, soy, fruits, etc. These feedstock are abundant in sugars, mainly sucrose, glucose, and fructose, and can be a promising strategy for BC production (see Table 2.2 in Section 2.3.2).

The presence of a complex medium in conventional feedstock can favor BC production, considering that the mixture between carbohydrates and other compounds can present an ideal combination to increase yield, as suggested by Premjet et al. (2007), who evaluated the association of nutrient sources that compose sugarcane molasses. By using conventional feedstock extracted from fruits (banana, pineapple, and strawberry) as a carbon source to produce BC using *K. rhaeticus*, He et al. (2020) observed that the process performance was improved compared to using HS medium (with glucose), suggesting the presence of elements in alternative carbon sources that performed a key role in increasing the process yield.

Besides being considered a nutrient source strategy, the average cost of BC production can be reduced. Machado et al. (2018) observed that by adding sugarcane molasses to the BC production there was a 20.06% reduction compared to the HS medium, suggesting that the alternative feedstock to the synthetic medium may be a viable strategy for large-scale implementation.

To maximize BC production and reduce production costs, enabling diversification of the applications of this versatile bioproduct, alternatives to synthetic culture media are of great importance. Therefore, advances in research aiming to evaluate the behavior of microorganisms producing BC in synthetic medium with various carbon sources and by total or partial replacement by conventional feedstock have a key role in the advances in BC production.

2.3 BACTERIAL CELLULOSE: FROM MICROORGANISMS TO PRODUCTION

2.3.1 Synthetic Medium

BC production performed in synthetic medium is frequently conducted using HS medium (Hestrin and Schramm 1954). Glucose is used as the carbon source in this medium, but several researchers have investigated modifications of the carbon source to evaluate the performance of the microorganism and BC synthesis (Table 2.1). The carbon source is the main precursor of BC biosynthesis, and even if it is not possible to conduct a process with a carbon source as the only strategy to improve yield, it is one of the factors that most affect the parameters and quality of BC. Glucose, fructose, sucrose, lactose, mannitol, galactose, among other compounds, are often used in synthetic media by researchers mainly aiming to understand the dynamics of different strains of bacteria that produce BC, and to enable the replacement of synthetic media by conventional feedstock.

As previously discussed, the applications that are found for the bioproduct are highly dependent on the process, which directly affects the characteristics of the biomaterial. Nutrient concentrations in the medium are crucial factors for bioprocess performance and, in addition to affecting yield, can alter the structural organization of BC. Bagewadi et al. (2020) reported that the BC synthesized by *Bacillus licheniformis* ZBT2 from lactose as a carbon source presented improved water retention capacity after optimization by statistical tools of the culture medium, that is, for the same composition with different concentrations there was a change in the structural organization of the biomaterial produced.

The exploration of synthetic medium with carbon sources other than glucose has two main objectives. First, researchers aim to substitute glucose with more complex sources, such as sources with an abundant mixture of sugars. Second, the use of glucose is associated with the induction of gluconic acid formation in the culture medium, which is responsible for a significant reduction in pH and may result in yield decreases (Cielecka et al. 2021; Singhania et al. 2022; Semjonovs et al. 2017; Bagewadi et al. 2020). The microorganisms applied, the carbon source used, and the pH profile are interrelated factors and must be considered when evaluating BC production (Huang et al. 2022).

Lactose was suggested as a carbon source for BC production by *Bacillus licheniformis* ZBT2, since it did not directly lead to pH decline, and had the highest yield of BC production compared to glucose, fructose, sucrose, maltose, and galactose (Bagewadi et al. 2020). For *K. hansenii* SI1, BC synthesis was observed for a range of carbon sources in the study by Cielecka et al. (2021), among them fructose, galactose, glycerol, lactose, maltose, mannitol, and sucrose. The better performance was observed for glycerol, that was attributed to the metabolic performance compared to glucose; besides, sucrose was not efficiently metabolized by the strain, suggesting the absence of enzymes responsible for the hydrolysis of the disaccharide structure (Cielecka et al. 2021). Recently, *Lactobacillus plantarum* was evaluated for different carbon sources (glucose, fructose, mannitol, xylose, galactose, sucrose, starch), and performed better on glucose, followed by mannitol. Interesting results were also observed for sucrose and starch (Saleh et al. 2022). *K. xylinus*, commonly used in BC production, generally performs better in glucose-rich medium (Du et al. 2018; Fernández et al. 2019). Glucose was also the most interesting carbon source for BC production in co-culture of isolated kombucha bacteria (*K. saccharivorans LN886705, Brettanomyces bruxellensis MH393498,* and *Brettanomyces anomalus KY103303*), since it presented the highest yield compared to sucrose, lactose, and maltose (Avcioglu et al. 2021). These studies briefly discussed are presented in Table 2.1 in terms of BC production yield.

Nitrogen sources are also relevant factors for the system. Although nitrogen does not directly contribute to the BC production, it is fundamental for the metabolism of the synthesizing microorganisms (Kurosumi et al. 2009; Cielecka et al. 2021; Bagewadi et al. 2020). In recent studies it was

TABLE 2.1
Carbon Sources in Synthetic Media for Bacterial Cellulose (BC) Production

		Bacterial Cellulose Production					
Microorganisms	Isolation Site	Carbon Source	Carbon Concentration (g/L)	Culture Medium	Conditions	Bacterial Cellulose (g/L)	Reference
Kamagataeibacter/ intermedius JF2	Wine vinegar	Mannitol and glucose	30	Hestrin-Schramm (HS) medium modified (g L^{-1}): 5 peptone, 5 yeast extract, 2.7 Na_2HPO_4, 1.15 citric acid, 10 glucose, 20 mannitol	Static culture, 28°C, 6 d, 40 mL of medium in petri dishes (10 cm); initial inoculum 0.6 OD_{600nm}	1.6	(Fernández et al. 2019)
Kamagataeibacter xylinus 7351T	Culture collection	Glucose	10	HS medium (g L^{-1}): 5 peptone, 5 yeast extract, 2.7 Na_2HPO_4, 1.15 citric acid, 10 glucose	Static culture, 28°C, 9 d, 40 mL of medium in petri dishes (10 cm); initial inoculum 0.6 OD_{600nm}	0.65	(Fernández et al. 2019)
Kamagataeibacter xylinus TJU-S8	Chinese persimomon vinegar	Glucose	19,14	HS medium modified (g L^{-1}): 19,142 glucose, 7 peptone, 10 yeast extract, 2.7 Na_2HPO_4, 0.5 citric acid, 35 ethanol, pH 6.03	Static culture, 30°C, 18 h, 250 mL of medium in Erlenmeyer flask (500 mL); initial inoculum 2% (v v^{-1})	4.62	(Du et al. 2018)
Bacillus licheniformis ZBT2	Forest soil	Galactose	20	HS medium modified (g L^{-1}): 20 galactose, 5 yeast extract; 5 peptone, 1.15 citrate, 2.7 Na_2HPO_4, 0.1 $MgSO_4.7H_2O$, 0.2 $CaCl_2$, 1.0 NaCl, pH 6.0	Static culture, 50°C, 3 d, 100 mL of medium in Erlenmeyer flask (250 mL); initial inoculum 10% (v v^{-1})	2.5	(Bagewadi et al. 2020)
		Lactose	20	20 g L^{-1} lactose, 13 g L^{-1} yeast extract, 1 g L^{-1} $MgSO_4$, 4 g L^{-1} KH_2PO_4, 7 ml L^{-1} ethanol, pH 7.2		9.2	
Komagataeibacter rhaeticus P1463	Kombucha	Fructose	20	HS medium (g L^{-1}): 5 peptone, 5 yeast extract, 2.7 Na_2HPO_4, 1.15 citric acid, 20 fructose	Static culture, 30°C, 5 d, 50 mL of medium in Erlenmeyer flask (300 mL); initial inoculum 5% (v v^{-1})	4.40	(Semjonovs et al. 2017)
		Glucose	20	HS medium (g L^{-1}): 5 peptone, 5 yeast extract, 2.7 Na_2HPO_4, 1.15 citric acid, 20 glucose		3.34	
Komagataeibacter hansenii B22	Kombucha	Fructose	20	HS medium (g L^{-1}): 5 peptone, 5 yeast extract, 2.7 Na_2HPO_4, 1.15 citric acid, 20 fructose	Static culture, 30°C, 5 d, 50 mL of medium in Erlenmeyer flask (300 mL); initial inoculum 5% (v v^{-1})	3.30	(Semjonovs et al. 2017)
		Glucose	20	HS medium (g L^{-1}): 5 peptone, 5 yeast extract, 2.7 Na_2HPO_4, 1.15 citric acid, 20 glucose		1.80	

Lactiplantibacillus plantarum	Rotten fruits	Glucose	15	15 g L^{-1} glucose, 13 g L^{-1} yeast extract, 1 g L^{-1} $MgSO_4$, 4 g L^{-1} KH_2PO_4, 7 ml L^{-1} ethanol, pH 7.2	Static culture, 30°C, 8 d, 50 mL of medium in Erlenmeyer flask (250 mL); initial inoculum 11% (v v^{-1})	4.51	(Saleh et al. 2022)
Lactiplantibacillus plantarum	Rotten fruits	Mannitol	20	Yamanaka medium modified (g L^{-1}): 20 glucose, 0.5 yeast extract, 0.5 $(NH_4)_2SO_4$, 0.3 KH_2PO_4, 0.005 $MgSO_4.7H_2O$	Static culture, 30°C, 7 d, 50 mL of medium in Erlenmeyer flask (250 mL); initial inoculum 10% (v v^{-1})	3.4	(Saleh et al. 2022)
Komagataeibacter saccharivorans LN886705 *Brettanomyces bruxellensis* MH393498 *Brettanomyces anomalus* KY103303	Kombucha	Glucose	60	60 g L^{-1} glucose, 10 g L^{-1} black tea. pH 6.0	Static culture, 30°C, 10 d; initial inoculum 10% (v v^{-1})	18.68	(Avcioglu et al. 2021)
Kamagataeibacter (Gluconacetobacter) xylinus TJU-D2	Vinegar	Glucose	19,58	19.58 g.L^{-1} glucose, 7 g.L^{-1} tryptone, 10 g.L^{-1} yeast extract, 3.85 g.L^{-1} ethanol, 0.5 g.L^{-1} citric acid, pH 6.32	Static culture, 30°C, 8 d, 300 mL of medium in Erlenmeyer flask (500 mL); initial inoculum 2% (v.v^{-1})	4.82	(Du et al. 2018)
Komagataeibacter medellinensis	Vinegar	Glucose	17,5	HS medium (g L^{-1}): 5 peptone, 20 yeast extract, 2.7 Na_2HPO_4, 20 glucose, citric acid until pH 3.2	Static culture, 30°C, 15 d; initial inoculum 10% (v.v^{-1})	3.3	(Molina-Ramírez et al. 2017)
Bacillus velezensis	Seawater	Glucose	20	HS medium (g L^{-1}): 5 peptone, 5 yeast extract, 2.7 Na_2HPO_4, 1.15 citric acid, 20 glucose, pH 5.0	Static culture, 30°C, 10 d, 50 mL of medium in Erlenmeyer flask (100 mL); initial inoculum 2% (v.v^{-1})	5.2	(Abouelkheir et al. 2020)
Komagataeibacter rhaeticus PG2	Rotten pomegranate	Glycerol	30	HS medium (g L^{-1}): 5 peptone, 5 yeast extract, 2.7 Na_2HPO_4, 1.15 citric acid, 30 glycerol, pH 6.0	Static culture, 28°C, 15 d, 100 mL of medium in Erlenmeyer flask (250 mL)	8.7	(Thorat and Dastager 2018)
Komagataeibacter rhaeticus PG2	Rotten pomegranate	Glucose	20	HS medium (g L^{-1}): 5 peptone, 5 yeast extract, 2.7 Na_2HPO_4, 1.15 citric acid, 20 glucose, pH 6.0	Static culture, 28°C, 15 d, 100 mL of medium in Erlenmeyer flask (250 mL)	4.05	(Thorat and Dastager 2018)

reported that the absence of nitrogen sources inhibits the synthesis of BC (Kurosumi et al. 2009), and that optimized nitrogen compounds at different concentrations can significantly improve yields (Saleh et al. 2022). Therefore, adequate nitrogen supplementation affects the BC yield, and the impact of different organic and inorganic nitrogen sources have already been investigated for different microorganisms. The relevance of the nitrogen and other nutrient sources should be considered also in processes where complex media composed of conventional raw feedstock (molasses and juices) are used, as can be seen in Table 2.2, which shows that most studies supplement the media with nitrogen and other nutrient sources to achieve satisfactory yields.

According to recent studies detailed in Tables 2.1 and 2.2, the environmental factors that most affect BC production are the optimum temperature, which will vary for each microorganism used, but generally close to 30°C; the pH, which for the HS medium is adjusted at 6.0, and, as already discussed, is crucial in the process, since depending on the carbon source it may suffer variations and will negatively affect the yield; the supply of oxygen dissolved in the culture medium, considering that the microorganisms that synthesize BC do so via an aerobic metabolic pathway, thus an inadequate supply of oxygen may inactivate the inoculum (Kadier et al. 2021).

Related to oxygen dissolution, BC production can be conducted in static culture and is the most described on a laboratory scale (see Tables 2.1 and 2.3). The static culture process is conducted by oxygen deprivation in the culture medium, which causes the dispersing microorganisms to move to the surface where the cells multiply and synthesize the biopolymer, resulting in the formation of cellulose microfibrils at the air-liquid interface of the bioreactor, and these microfibrils group together to form a cellulose film, which has a uniform and smooth structure when compared to the bioproducts of an agitated culture process (Raghavendran et al. 2020; Singhsa et al. 2018).

In agitated cultures, bioreactors or flasks with boosters or aeration are used to improve oxygen transfer and cell dispersion. The cellulose film is formed dispersed in the culture medium in the form of irregular pellets (Raghavendran et al. 2020; Singhania et al. 2022; Singhsa et al. 2018; Gao et al. 2020). The processes can still be conducted with semi-continuous aeration (Krusong et al. 2021). It is important to note that each process (agitated or static) can change the performance of the strains and the morphological, physical, and mechanical properties of the formed BC, and influence the application potential of the bioproduct.

To establish a high-quality commercial BC production from a feedstock rich in sugars and of cheap acquisition cost, the optimization of the cultivation conditions is a good strategy before using more complex media, aiming at the development of a robust bioprocess that results in a high technology product (Singhania et al. 2022). In this perspective, approaches using statistical tools are widely applied for the optimization of conditions and can be crucial for understanding the dynamics of the bioprocess of BC production, which can increase yield, reduce cultivation time, and minimize costs. These approaches have been corroborated by some recent studies. Du et al. (2018) evaluated experimentally the optimization strategy with respect to the composition of the culture medium (glucose, yeast extract, peptone, Na_2HPO_4, citric acid, ethanol, temperature, and pH) with *K. xylinus* bacteria, and achieved a maximum yield of BC in 18 hours of cultivation 1.46 times higher compared to the non-optimized medium (Table 2.1). Bagewadi et al. (2020) also applied experimental optimization strategies, achieving a maximum yield result of 6.6 times higher than the non-optimized medium. However, although synthetic culture media are the most described for BC production, their use in industrial applications is limited, mainly due to the cost. Therefore, research is being conducted into alternatives that include media conventionally used in industry, and which are discussed in the following section.

2.3.2 Conventional Feedstock for Culture Medium

BC production using conventional feedstock abundant in sugars is a challenge that has been widely studied to improve yields and reduce production costs, considering that most of these provide the necessary nutrients to support microorganism growth and BC biosynthesis (Table 2.2). Besides

TABLE 2.2
Bacterial Cellulose (BC) Production Research That Used Conventional Feedstock as an Alternative Carbon Source

Conventional Feedstock		Bacterial Cellulose Production				
Carbon Source	Composition	Initial Carbon Concentration (g/L)	Supplement Sources	Conditions	Bacterial Cellulose (g L^{-1})	Reference
Sugarcane molasses	Sucrose (57.8%), glucose (5.2%), fructose (6.5%) and other noncarbohydrate substances (15.8%)	50	4 $g.L^{-1}$ yeast extract, 20 $g.L^{-1}$ ethanol	Static culture, 30°C, 5 d, 45 mL of medium in Erlenmeyer flask (250 mL); initial inoculum 1:10 ($v\ v^{-1}$)	1.9	(Machado et al. 2018)
Sugarcane juice	Sucrose (19.15%), glucose (2.42%), fructose (3.85%)	263.7	-	Static culture, 30°C, 10 d, pH 4.5; initial inoculum 10^6 UFC mL^{-1}	0.8	(Jaroennonthasit et al. 2021)
Coconut juice	Sucrose (7.08%), glucose (1.60%), fructose (1.81%)	108.3	-	Static culture, 30°C, 10 d, pH 4.5; initial inoculum 10^6 UFC mL^{-1}	0.33	(Jaroennonthasit et al. 2021)
Apple juice	Sucrose (18.1%), glucose (12.1%), fructose (49.9%)	20	5 $g.L^{-1}$ peptone, 5 $g.L^{-1}$ yeast extract, 2.7 $g.L^{-1}$ Na_2HPO_4, 1.15 $g.L^{-1}$ citric acid	Static culture, 30°C, 5 d, 50 mL of medium in Erlenmeyer flask (300 mL); initial inoculum 5% ($v\ v^{-1}$)	3.5	(Semjonovs et al. 2017)
Apple juice	Sucrose (18.1%), glucose (12.1%), fructose (49.9%)	20	5 $g.L^{-1}$ peptone, 5 $g.L^{-1}$ yeast extract, 2.7 $g.L^{-1}$ Na_2HPO_4, 1.15 $g.L^{-1}$ citric acid	Static culture, 30°C, 5 d, 50 mL of medium in Erlenmeyer flask (300 mL); initial inoculum 5% ($v\ v^{-1}$)	4.77	(Semjonovs et al. 2017)
Mango extract (pulp juice)	Reducing sugar 16.9% (measured as glucose)	-	20 $g.L^{-1}$ yeast extract, 3% ($v.v^{-1}$) ethanol, acetate buffer (100 mmol L^{-1}, pH 4.7)	Static culture, 30°C, 21 d	25.3	(Calderón-Toledo et al. 2022)
Chinese wolfberry juice	-	25.65	-	Static culture, 30°C, 16 d, 200 mL of medium in Erlenmeyer flask (500 mL)	3.145	(Zhang et al. 2021)
Orange pulp	Sucrose (4.8%), glucose (1.3%), fructose (1.2%), pH 3.9	73	20 $g.L^{-1}$ peptone, 5 $g.L^{-1}$ yeast extract, 1.2 $g.L^{-1}$ citric acid	Static culture, 30°C, 14 d, pH 6.0	5.9	(Kurosumi et al. 2009)
Pineapple juice	Sucrose (4.9%), glucose (2.0%), fructose (2.0%), pH 3.5	89	20 $g.L^{-1}$ peptone, 5 $g.L^{-1}$ yeast extract, 1.2 $g.L^{-1}$ citric acid	Static culture, 30°C, 14 d, pH 6.0	4.1	(Kurosumi et al. 2009)
Grape juice	Sucrose (0.2%), glucose (4.2%), fructose (5.9%), pH 4.1	103	20 $g.L^{-1}$ peptone, 5 $g.L^{-1}$ yeast extract, 1.2 $g.L^{-1}$ citric acid	Static culture, 30°C, 14 d, pH 6.0	1.8	(Kurosumi et al. 2009)

(Continued)

TABLE 2.2 *(Continued)*

Bacterial Cellulose (BC) Production Research That Used Conventional Feedstock as an Alternative Carbon Source

Conventional Feedstock		Bacterial Cellulose Production				
Carbon Source	Composition	Initial Carbon Concentration (g/L)	Supplement Sources	Conditions	Bacterial Cellulose (g L^{-1})	Reference
Pineapple juice	-	20	5 g.L^{-1} peptone, 5 g.L^{-1} yeast extract, 1.2 g.L^{-1} citric acid, 2.7 g.L^{-1} Na_2HPO_4 (pH 6.0)	Static culture, 30°C, 10 d, 100 mL of medium in Erlenmeyer flask (250 mL); initial inoculum 10% (v v^{-1})	3.8	(He et al. 2020)
Strawberry juice	-	20	5 g.L^{-1} peptone, 5 g.L^{-1} yeast extract, 1.2 g.L^{-1} citric acid, 2.7 g.L^{-1} Na_2HPO_4 (pH 6.0)	Static culture, 30°C, 10 d, 100 mL of medium in Erlenmeyer flask (250 mL); initial inoculum 10% (v v^{-1})	3.3	(He et al. 2020)
Banana juice	-	20	5 g.L^{-1} peptone, 5 g.L^{-1} yeast extract, 1.2 g.L^{-1} citric acid, 2.7 g.L^{-1} Na_2HPO_4 (pH 6.0)	Static culture, 30°C, 10 d, 100 mL of medium in Erlenmeyer flask (250 mL); initial inoculum 10% (v v^{-1})	3.1	(He et al. 2020)
Vegetable juice	-	20	5 g.L^{-1} peptone, 5 g.L^{-1} yeast extract, 1.2 g.L^{-1} citric acid, 2.7 g.L^{-1} Na_2HPO_4 (pH 6.0)	Static culture, 30°C, 10 d, 100 mL of medium in Erlenmeyer flask (250 mL); initial inoculum 10% (v v^{-1})	2.4	(He et al. 2020)

being a medium rich in naturally available sugars, conventional feedstock do not require complex and costly upstream processes such as chemical pretreatments or hydrolysis steps (Singhania et al. 2022; Gorgieva and Trček 2019).

As shown in Table 2.2, several carbon sources have already been used for BC biosynthesis and have enabled the expansion of applications beyond processes that require strict quality control, such as in the medical field. A recent study conducted by Huang et al. (2022) utilized sugarcane molasses as carbon source (substituting glucose in HS) for *K. xylinus*. The process was performed in alkaline medium (BG-11) and aimed at the immobilization of microalgae *Chlorella sorokiniana* AK-1, for application in wastewater treatment. The results demonstrated satisfactory BC production in 50:50 culture medium (50% HS-molasses medium; 50% BG11 medium) (Huang et al. 2022). He et al. (2020) used different carbon sources of conventional feedstock for the *K. rhaeticus* TJPU03 and observed a higher yield in fruit juices (banana, pineapple, and strawberry) compared to the control (HS medium), demonstrating the biotechnological potential of investigations in this field.

In using complex culture medium, with alternative carbon or nitrogen sources to pure synthetic sources, an evaluation of the structural and physicochemical properties of the BC produced should be considered to ensure that it is suitable for the desired application (Huang et al. 2022). Many studies report that using carbon sources from conventional media does not significantly change the physical characteristics of BC compared to synthetic medium; however, the physical properties of the structure can be affected by changing the feedstock (Singhsa et al. 2018; Abol-Fotouh et al. 2020; Machado et al. 2018). It has been reported that changes in the physical characteristics observed in assays, for example, scanning electron microscopy (SEM), may be associated with carbon sources with widely varying sugar concentrations and other compounds, which can disrupt the arrangement of the matrix of the BC (Jaroennonthasit et al. 2021).

Abol-Fotouh et al. (2020) performed a study with the *K. saccharivorans* MD1 strain on extract of palm date fruits, fig fruits, and sugarcane molasses, and no microstructural and morphological variations were observed in nanofiber dimensions or network branching, which ranged from 10 to 90 nm similarly in all culture mediums. For *K. rhaeticus* SEM images were performed for synthetic medium (HS) and sugarcane molasses medium, and it was observed that the samples presented a porous structure composed of cellulose nanofibers of different average thickness between the synthetic and conventional mediums, being 92 nm for HS and 119 nm for sugarcane molasses, that is, the culture medium composed of sugarcane molasses induced the formation of a thicker cellulose structure (Machado et al. 2018). These results corroborate those of Jaroennonthasit et al. (2021), who evaluated *K. xylinus* in four different media (glucose, sugarcane juice, molasses, and coconut juice) and observed that the carbon source affected the arrangement of the BC matrix, with the presence of high concentrations of sucrose in the sugarcane-based medium resulting in an irregular layering of the microfibrils of the matrix. A similar result was obtained by Çakar et al. (2014), who observed denser microfibrils biosynthesized by *K. xylinus* FC01 in molasses medium compared to a thinner microfibril in HS medium.

The physical properties of the structure can be altered with alternatives to synthetic carbon sources. The thermal stability of BC can be investigated with regard to the thermal transition, melting point, and decomposition, which are relevant characteristics for commercial application (Rahman et al. 2021). By thermogravimetric analysis of BC synthesized by *Acetobacter pasteurianus* RSV-4 from tomato juice it was found to have a maximum weight loss rate at approximately 343°C, and good thermal stability at approximately 200°C with a mass loss of less than 80% (Kumar et al. 2019).

Characterization of the BC structures is crucial for commercial applications. X-ray diffraction tests, thermogravimetry, mechanical properties, water retention capacity, porosity analysis, biocompatibility, and cytotoxicity are some of the assays that can be applied to evaluate the synthesized biomaterial, being important when replacing the carbon source with conventional feedstock, mainly considering the complexity of the medium and the possibility of modifications in the biosynthesis processes, which can expand or limit the applications of the bioproduct. The availability of different feedstock enables a variety of applications to be targeted for BC, which can be a potential strategy for scale-up and commercialization.

2.4 PERSPECTIVES AND CHALLENGES OF BACTERIAL CELLULOSE PRODUCTION

Bacterial cellulose can be produced basically by three different culture methods: static, agitated/shaking, and bioreactor culture. While the static culture will result in the accumulation of a cellulose gelatinous pellicle at the surface of the culture, the agitated culture will result in cellulose pellets or irregular masses. The use of these different methods will result in different BC properties, morphology, and structure (Wang et al. 2019).

Bacterial cellulose produced by the static culture method has great advantages, such as good tensile strength, water permeability, tearing resistance, binding capacity, elastic modulus, and a compact texture. Due to its great biocompatibility and biodegradability, it can be applied in different fields, especially in the medical field (Andriani et al. 2020). However, static culture provides a low productivity due to the long time needed for cultivation and the extensive space needed for the BC production (Hsieh et al. 2016).

Due to the low productivity presented by static culture, agitated fermentation is the most used culture method for BC production for commercial purposes. The pellets, spheres of irregular masses of BC formed by the agitated culture have lower DP, crystallinity, and mechanical strength compared to the membrane formed in static culture. The mechanism of formation of these irregular masses remains unknown (Esa et al. 2014). However, it is known that the altered microfibril structure is due to the disruption effect on the formation of hydrogen bonds in the cellulose caused by the aeration (Bootten et al. 2008). The main disadvantage of using agitated culture is the action of cellulose-negative mutants (Cel−), which can grow faster than cellulose-producing cells (Cel+). The shear stress in agitated culture generates Cel−, favoring the production of byproducts and, consequently, decreasing the cellulose yield (Campano et al. 2015). In fact, Krusong et al. (2021) produced reduced amounts of BC using higher aeration rates, which was accompanied by the reduction of total cell mass. According to the authors, this effect is most likely due to a gradual replacement of Cel+ cells by Cel− cells or due to the spontaneous appearance of Cel− cells.

To overcome the problem of Cel− in BC production, different bioreactor systems have been tested, such as using airlift, rotating disks, or biofilm support (Andriani et al. 2020). Table 2.3 shows a summary of the mains advantages and disadvantages of each method.

TABLE 2.3
Summary of Culture Methods for Bacterial Cellulose (BC) Production and Their Main Advantages and Disadvantages

Culture Method	Advantages	Disadvantages	Reference
Static	Good tensile strength, water permeability, tearing resistance, binding capacity, elastic modulus, compact texture	Long production time, extensive space needed for production, high production cost	(Andriani et al. 2020; Hsieh et al. 2016)
Agitated	Greater productivity when compared to the static culture, better oxygen delivery, less space needed	Lower degree of polymerization, crystallinity, and mechanical strength; generation of Cel− cells	(Esa et al. 2014)
Air-lift bioreactor	Low power consumption, simple mechanical structure, high oxygen transfer	Weak mechanical strength; the adhesion of BC to the inside wall and the upper part of the apparatus can be a problem	(Shoda and Sugano 2005; Andriani et al. 2020)

Culture Method	Advantages	Disadvantages	Reference
Rotating disk bioreactor	Production of homogeneous cellulose; overcomes the problem of the adhesion of BC in the apparatus; higher tensile strength	BC yield is lower compared to the static process	(Gorgieva and Trček 2019; Andriani et al. 2020; Shoda and Sugano 2005)
Biofilm reactor	Higher BC production	Decreased crystallinity and crystal size	(Cheng et al. 2011)

2.5 TECHNOLOGY AND INNOVATION IN PRODUCTION OF BACTERIAL CELLULOSE FROM CONVENTIONAL FEEDSTOCK

Several patents can be found on the production of BC using conventional sugars, such as glucose and saccharose, but few specify the feedstock from which each of these sugars are derived. The patent BR102014027203A2 (Melo and Aguiar 2014) specifies the production of BC by *Gluconoacetobacter hansenii* using sugar from renewable sources such as sugarcane juice, molasses, or its syrup. To produce the BC, the inventors used biodigesters having a flotation surface four times the height of the apparatus. According to the inventors, with this configuration, they were able to increase the conversion rate of the sugars from renewable source to polymer mass. In the patent CN113355377A (Hong et al. 2021) the inventors were able to produce a BC facial mask using both glucose and granulated sugar (saccharose) using *Acetobacter gluconicum*. Furthermore, the facial mask also contained skin probiotic lactic acid bacteria.

The patent JP2021119747A claims a scaffolding material for cell culture that contains BC was produced by the bacterium *Gluconacetobacter intermedius*. The production was conducted using carbon sources such as monosaccharides (glucose and fructose), disaccharides (such as sucrose, maltose, and lactose), and oligosaccharides. The inventors specify that these carbon sources could come from conventional feedstock such as sugarcane, sugar beet, sugar palm, and sweet sorghum. The scaffolding material can be used, for example, in the medical field for the cultivation of stem cells (Ishida et al. 2020).

Different technologies and innovations were studied to overcome the problems of BC production. The patent US5955326A (Bungay and Serafica 1995) pioneered the use of a rotating disk bioreactor, which can overcome the problem of production of Cel– cells and, according to the inventors, it produced a coherent gel-like pellicle desirable for many uses, such as a wound dressing. In the patent CN102533904A (Hong et al. 2012), the inventors used a rotating drum bioreactor. According to them, the use of this bioreactor is a way to overcome the low production efficiency and the large area needed to produce BC using a static culture.

Although the first report of BC was more than a century ago, the subject is still of interest, as evidenced by the increase in the number of patents in the past decades (Bianchet et al. 2020; Charreau et al. 2012). That is to be expected because of the growing interest in materials that are renewable and biodegradable, and with properties, such as light weight, high mechanical properties, and high chemical purity. Based on that, the number of innovative studies on BC is expected to grow worldwide.

2.6 CONCLUSIONS

Bacterial cellulose is a biopolymer with great characteristics such as high structural crystallinity, high water retention capacity, elasticity, tensile strength, and biocompatibility. These aspects are even more significant with the increasing interest in materials that are renewable and biodegradable, both attributes present in BC. This polymer can be produced using different types of sugars that can proceed from conventional feedstock, such as sugarcane, corn, and beetroot. In fact, the use of different nutrient sources can directly affect the characteristics and quality of BC, also affecting its future application. One of the weaknesses of BC is its production cost. Because of that the search for cheaper feedstock is of utmost importance since it can directly affect the final cost of the product.

2.7 REFERENCES

Abol-Fotouh, Deyaa, Mohamed A. Hassan, Hassan Shokry, Anna Roig, Mohamed S. Azab, and Abd El-Hady B. Kashyout. 2020. "Bacterial Nanocellulose from Agro-Industrial Wastes: Low-Cost and Enhanced Production by *Komagataeibacter saccharivorans* MD1." *Scientific Reports* 10 (1): 3491. doi:10.1038/s41598-020-60315-9.

Abouelkheir, Samia S., Marwa S. Kamara, Salma M. Atia, Sara A. Amer, Marina I. Youssef, Rana S. Abdelkawy, Sherine N. Khattab, and Soraya A. Sabry. 2020. "Novel Research on Nanocellulose Production by a Marine *Bacillus velezensis* Strain SMR: A Comparative Study." *Scientific Reports* 10 (1): 14202. doi:10.1038/s41598-020-70857-7.

Andriani, Dian, Arina Yuthi Apriyana, and Myrtha Karina. 2020. "The Optimization of Bacterial Cellulose Production and Its Applications: A Review." *Cellulose* 27 (12). Springer: 6747–6766. doi:10.1007/S10570-020-03273-9.

Avcioglu, Nermin Hande, Meric Birben, and Isil Seyis Bilkay. 2021. "Optimization and Physicochemical Characterization of Enhanced Microbial Cellulose Production with a New Kombucha Consortium." *Process Biochemistry* 108 (September): 60–68. doi:10.1016/j.procbio.2021.06.005.

Bagewadi, Zabin K., Jayateerth S. Bhavikatti, Uday M. Muddapur, Deepak A. Yaraguppi, and Sikandar I. Mulla. 2020. "Statistical Optimization and Characterization of Bacterial Cellulose Produced by Isolated Thermophilic *Bacillus licheniformis* Strain ZBT2." *Carbohydrate Research* 491 (May): 107979. doi:10.1016/j.carres.2020.107979.

Bianchet, Ritanara Tayane, Anelise Leal Vieira Cubas, Marina Medeiros Machado, and Elisa Helena Siegel Moecke. 2020. "Applicability of Bacterial Cellulose in Cosmetics: Bibliometric Review." *Biotechnology Reports* 27 (September). Elsevier: e00502. doi:10.1016/J.BTRE.2020.E00502.

Bootten, Tracey J., Philip J. Harris, Laurence D. Melton, and Roger H. Newman. 2008. "WAXS and 13C NMR Study of *Gluconoacetobacter xylinus* Cellulose in Composites with Tamarind Xyloglucan." *Carbohydrate Research* 343 (2). Elsevier: 221–229. doi:10.1016/J.CARRES.2007.11.008.

Brown, Adrian J. 1886. "XLIII: On an Acetic Ferment Which Forms Cellulose." *J. Chem. Soc., Trans.* 49: 432–439. doi:10.1039/CT8864900432.

Bungay, H. R., and G. C. Serafica. 1995. "Production of Microbial Cellulose Using a Rotating Disk Film Bioreactor." USA.

Çakar, Fatih, Işılay Özer, A. Özhan Aytekin, and Fikrettin Şahin. 2014. "Improvement Production of Bacterial Cellulose by Semi-Continuous Process in Molasses Medium." *Carbohydrate Polymers* 106 (June): 7–13. doi:10.1016/j.carbpol.2014.01.103.

Calderón-Toledo, Susana, Manuel Horue, Vera A. Alvarez, Guillermo R. Castro, and Amparo Iris Zavaleta. 2022. "Isolation and Partial Characterization of *Komagataeibacter* sp. SU12 and Optimization of Bacterial Cellulose Production Using *Mangifera indica* Extracts." *Journal of Chemical Technology & Biotechnology* 97 (6): 1482–1493. doi:10.1002/jctb.6839.

Campano, Cristina, Ana Balea, Angeles Blanco, and Carlos Negro. 2015. "Enhancement of the Fermentation Process and Properties of Bacterial Cellulose: A Review." *Cellulose* 23 (1). Springer: 57–91. doi:10.1007/S10570-015-0802-0.

Charreau, Hernan, Maria L. Foresti, and Analia Vazquez. 2012. "Nanocellulose Patents Trends: A Comprehensive Review on Patents on Cellulose Nanocrystals, Microfibrillated and Bacterial Cellulose." *Recent Patents on Nanotechnology* 7 (1). Bentham Science Publishers Ltd.: 56–80. doi:10.2174/187221013804484854.

Cheng, Kuan Chen, Jeffrey M. Catchmark, and Ali Demirci. 2011. "Effects of CMC Addition on Bacterial Cellulose Production in a Biofilm Reactor and Its Paper Sheets Analysis." *Biomacromolecules* 12 (3). American Chemical Society: 730–736. doi:10.1021/BM101363T/SUPPL_FILE/BM101363T_SI_001.PDF.

Cielecka, Izabela, Małgorzata Ryngajłło, Waldemar Maniukiewicz, and Stanisław Bielecki. 2021. "Highly Stretchable Bacterial Cellulose Produced by *Komagataeibacter hansenii* SI1." *Polymers* 13 (24): 4455. doi:10.3390/polym13244455.

Du, Renpeng, Fangkun Zhao, Qian Peng, Zhijiang Zhou, and Ye Han. 2018. "Production and Characterization of Bacterial Cellulose Produced by *Gluconacetobacter xylinus* Isolated from Chinese Persimmon Vinegar." *Carbohydrate Polymers* 194 (August): 200–207. doi:10.1016/j.carbpol.2018.04.041.

Esa, Faezah, Siti Masrinda Tasirin, and Norliza Abd Rahman. 2014. "Overview of Bacterial Cellulose Production and Application." *Agriculture and Agricultural Science Procedia* 2 (January). Elsevier: 113–119. doi:10.1016/J.AASPRO.2014.11.017.

Fernández, Julia, A. Gala Morena, Susana V. Valenzuela, F. I. Javier Pastor, Pilar Díaz, and Josefina Martínez. 2019. "Microbial Cellulose from a *Komagataeibacter intermedius* Strain Isolated from Commercial Wine Vinegar." *Journal of Polymers and the Environment* 27 (5): 956–967. doi:10.1007/s10924-019-01403-4.

Gao, Hongliang, Qian Sun, Zebei Han, Jiahe Li, Bowen Liao, Lulu Hu, Jie Huang, et al. 2020. "Comparison of Bacterial Nanocellulose Produced by Different Strains Under Static and Agitated Culture Conditions." *Carbohydrate Polymers* 227 (January): 115323. doi:10.1016/j.carbpol.2019.115323.

Gorgieva Selestina, and Trček Gorgieva. 2019. "Bacterial Cellulose: Production, Modification and Perspectives in Biomedical Applications." *Nanomaterials* 9 (10): 1352. doi:10.3390/nano9101352.

He, Xiaoling, Hongyan Meng, Haode Song, Shiyu Deng, Tingting He, Shuai Wang, Dongsheng Wei, and Zhijun Zhang. 2020. "Novel Bacterial Cellulose Membrane Biosynthesized by a New and Highly Efficient Producer *Komagataeibacter rhaeticus* TJPU03." *Carbohydrate Research* 493 (July): 108030. doi:10.1016/j.carres.2020.108030.

Hestrin, S., and M. Schramm. 1954. "Synthesis of Cellulose by *Acetobacter xylinum* II. Preparation of Freeze-Dried Cells Capable of Polymerizing Glucose to Cellulose." *Biochemical Journal* 58 (2): 345–352. doi:10.1042/bj0580345.

Hong, D. S., J. In, H. S. Kim, T. H. Kim, P. J. Ho, J. W. Choi, S. H. Lee, et al. 2021. "Mixed Bacteria Cellulose Facial Mask with Skin Microbial Population Regulating Effect and Preparation Method Thereof." China.

Hong, F., X. Yang, and S. Tang. 2012. "Method and Device for Preparing Bacterial Cellulose Composite Material Quickly on Large Scale." China.

Hsieh, Jung Ting, Meng Jiy Wang, Jinn Tsyy Lai, and Hwai Shen Liu. 2016. "A Novel Static Cultivation of Bacterial Cellulose Production by Intermittent Feeding Strategy." *Journal of the Taiwan Institute of Chemical Engineers* 63 (June). Elsevier: 46–51. doi:10.1016/J.JTICE.2016.03.020.

Huang, Wei-Ming, Jih-Heng Chen, Dillirani Nagarajan, Cheng-Kang Lee, Sunita Varjani, Duu-Jong Lee, and Jo-Shu Chang. 2022. "Immobilization of *Chlorella sorokiniana* AK-1 in Bacterial Cellulose by Co-Culture and Its Application in Wastewater Treatment." *Journal of the Taiwan Institute of Chemical Engineers* (March): 104286. doi:10.1016/j.jtice.2022.104286.

Ishida, T., H. Ando, T. Kusano, and T. Matsushima. 2020. "Cell Culture Scaffold Material." Japan.

Jaroennonthasit, Watcharapon, Nga Tien Lam, and Prakit Sukyai. 2021. "Evaluation of Carbon Sources from Sugar Industry to Bacterial Nanocellulose Produced by *Komagataeibacter xylinus*." *International Journal of Biological Macromolecules* 191 (November): 299–304. doi:10.1016/j.ijbiomac.2021.09.028.

Jozala, Angela Faustino, Leticia Celia de Lencastre-Novaes, André Moreni Lopes, Valéria de Carvalho Santos-Ebinuma, Priscila Gava Mazzola, Adalberto Pessoa-Jr, Denise Grotto, Marli Gerenutti, and Marco Vinicius Chaud. 2016. "Bacterial Nanocellulose Production and Application: A 10-Year Overview." *Applied Microbiology and Biotechnology* 100 (5): 2063–2072. doi:10.1007/s00253-015-7243-4.

Kadier, Abudukeremu, R. A. Ilyas, M. R. M. Huzaifah, Nani Harihastuti, S. M. Sapuan, M. M. Harussani, M. N. M. Azlin, et al. 2021. "Use of Industrial Wastes as Sustainable Nutrient Sources for Bacterial Cellulose (BC) Production: Mechanism, Advances, and Future Perspectives." *Polymers* 13 (19): 3365. doi:10.3390/polym13193365.

Kosseva, Maria R., Siyi Zhong, Mengmeng Li, Juyan Zhang, and Natasia A. S. Tjutju. 2020. "Biopolymers Produced from Food Wastes: A Case Study on Biosynthesis of Bacterial Cellulose from Fruit Juices." In *Food Industry Wastes*, edited by Maria R. Kosseva and Colin Webb, 2nd ed., 225–254. Elsevier. doi:10.1016/B978-0-12-817121-9.00011-5.

Krusong, Warawut, Ruttipron Pothimon, Salvatore La China, and Anthony Keith Thompson. 2021. "Consecutive Bacterial Cellulose Production by Luffa Sponge Enmeshed with Cellulose Microfibrils of *Acetobacter xylinum* under Continuous Aeration." *3 Biotech* 11 (1): 6. doi:10.1007/s13205-020-02569-8.

Kumar, Vinod, Devendra Kumar Sharma, Vasudha Bansal, Deepak Mehta, Rajender S. Sangwan, and Sudesh Kumar Yadav. 2019. "Efficient and Economic Process for the Production of Bacterial Cellulose from Isolated Strain of *Acetobacter pasteurianus* of RSV-4 Bacterium." *Bioresource Technology* 275 (March): 430–433. doi:10.1016/j.biortech.2018.12.042.

Kurosumi, Akihiro, Chizuru Sasaki, Yuya Yamashita, and Yoshitoshi Nakamura. 2009. "Utilization of Various Fruit Juices as Carbon Source for Production of Bacterial Cellulose by *Acetobacter xylinum* NBRC 13693." *Carbohydrate Polymers* 76 (2): 333–335. doi:10.1016/j.carbpol.2008.11.009.

Machado, Rachel T. A., Andréia Bagliotti Meneguin, Rafael Miguel Sábio, Douglas Faza Franco, Selma G. Antonio, Junkal Gutierrez, Agnieszka Tercjak, et al. 2018. "*Komagataeibacter rhaeticus* Grown in Sugarcane Molasses-Supplemented Culture Medium as a Strategy for Enhancing Bacterial Cellulose Production." *Industrial Crops and Products* 122 (October): 637–646. doi:10.1016/j.indcrop.2018.06.048.

Melo, F. A. D., and L. M. Aguiar. 2014. "Processo de Produção e Purificação Em Escala de Celulose Bacteriana Obtida Pela Polimerização Da Glicose a Partir de Açúcares de Fontes Renováveis via Biotecnologia Por Meio Da Propagação de *Gluconoacetobacter hansenii* LMSPE Em Reatores e Obtenção Da Celu." Brazil.

Molina-Ramírez, Carlos, Margarita Castro, Marlon Osorio, Mabel Torres-Taborda, Beatriz Gómez, Robin Zuluaga, Catalina Gómez, Piedad Gañán, Orlando Rojas, and Cristina Castro. 2017. "Effect of Different Carbon Sources on Bacterial Nanocellulose Production and Structure Using the Low PH Resistant Strain *Komagataeibacter medellinensis*." *Materials* 10 (6): 639. doi:10.3390/ma10060639.

Premjet, Siripong, Duangporn Premjet, and Yoshito Ohtani. 2007. "The Effect of Ingredients of Sugar Cane Molasses on Bacterial Cellulose Production by *Acetobacter xylinum* ATCC 10245." *Sen'i Gakkaishi* 63 (8): 193–199. doi:10.2115/fiber.63.193.

Raghavendran, Vijayendran, Emmanuel Asare, and Ipsita Roy. 2020. "Bacterial Cellulose: Biosynthesis, Production, and Applications." In *Advances in Microbial Physiology*, edited by Robert K. Poole, 89–138. Elsevier. doi:10.1016/bs.ampbs.2020.07.002.

Rahman, Sameeha Syed Abdul, T. Vaishnavi, G. Sai Vidyasri, K. Sathya, P. Priyanka, Ponnusami Venkatachalam, and Sugumaran Karuppiah. 2021. "Production of Bacterial Cellulose Using *Gluconacetobacter kombuchae* Immobilized on *Luffa aegyptiaca* Support." *Scientific Reports* 11 (1): 2912. doi:10.1038/s41598-021-82596-4.

Ross, P., R. Mayer, and M. Benziman. 1991. "Cellulose Biosynthesis and Function in Bacteria." *Microbiological Reviews* 55 (1): 35–58. doi:10.1128/mr.55.1.35-58.1991.

Saleh, Ahmed K., Hamada El-Gendi, Nadia A. Soliman, Waleed K. El-Zawawy, and Yasser R. Abdel-Fattah. 2022. "Bioprocess Development for Bacterial Cellulose Biosynthesis by Novel *Lactiplantibacillus plantarum* Isolate along with Characterization and Antimicrobial Assessment of Fabricated Membrane." *Scientific Reports* 12 (1): 2181. doi:10.1038/s41598-022-06117-7.

Semjonovs, Pavels, Maija Ruklisha, Longina Paegle, Madara Saka, Rita Treimane, Marite Skute, Linda Rozenberga, Laura Vikele, Martins Sabovics, and Ilse Cleenwerck. 2017. "Cellulose Synthesis by *Komagataeibacter rhaeticus* Strain P 1463 Isolated from Kombucha." *Applied Microbiology and Biotechnology* 101 (3): 1003–1012. doi:10.1007/s00253-016-7761-8.

Shoda, Makoto, and Yasushi Sugano. 2005. "Recent Advances in Bacterial Cellulose Production." *Biotechnology and Bioprocess Engineering* 10 (1). Springer: 1–8. doi:10.1007/BF02931175.

Singhania, Reeta Rani, Anil Kumar Patel, Yi-Sheng Tseng, Vinod Kumar, Chiu-Wen Chen, Dibyajyoti Haldar, Jitendra Kumar Saini, and Cheng-Di Dong. 2022. "Developments in Bioprocess for Bacterial Cellulose Production." *Bioresource Technology* 344 (January): 126343. doi:10.1016/j.biortech.2021.126343.

Singhsa, Pratyawadee, Ravin Narain, and Hathaikarn Manuspiya. 2018. "Physical Structure Variations of Bacterial Cellulose Produced by Different *Komagataeibacter xylinus* Strains and Carbon Sources in Static and Agitated Conditions." *Cellulose* 25 (3): 1571–1581. doi:10.1007/s10570-018-1699-1.

Thorat, Meghana N., and Syed G. Dastager. 2018. "High Yield Production of Cellulose by a *Komagataeibacter rhaeticus* PG2 Strain Isolated from Pomegranate as a New Host." *RSC Advances* 8 (52): 29797–29805. doi:10.1039/C8RA05295F.

Wang, Jing, Javad Tavakoli, and Youhong Tang. 2019. "Bacterial Cellulose Production, Properties and Applications with Different Culture Methods: A Review." *Carbohydrate Polymers* 219 (September). Elsevier: 63–76. doi:10.1016/J.CARBPOL.2019.05.008.

Yamada, Yuzo. 1983. "*Acetobacter xylinus* sp. Nov., Nom. Rev., for the Cellulose-Forming and Cellulose-Less, Acetate-Oxidizing Acetic Acid Bacteria with the Q-10 System." *The Journal of General and Applied Microbiology* 29 (5): 417–420. doi:10.2323/jgam.29.417.

Yamada, Yuzo, Pattaraporn Yukphan, Huong Thi Lan Vu, Yuki Muramatsu, Duangjai Ochaikul, and Yasuyoshi Nakagawa. 2012. "Subdivision of the Genus Gluconacetobacter Yamada, Hoshino and Ishikawa 1998: The Proposal of Komagatabacter Gen. Nov., for Strains Accommodated to the *Gluconacetobacter xylinus* Group in the α-Proteobacteria." *Annals of Microbiology* 62 (2): 849–859. doi:10.1007/s13213-011-0288-4.

Zhang, Tianzhen, Yuqing Shen, Senjia Zhang, Zexiong Xie, Xiyu Cheng, Wenchao Li, and Cheng Zhong. 2021. "Monosaccharide Removal and Effects of *Komagataeibacter xylinus* Fermentation on Antioxidant Capacity and Flavor Profile of Chinese Wolfberry Juice." *Journal of Food Processing and Preservation* 45 (10). doi:10.1111/jfpp.15800.

3 Bacterial Cellulose Production from Renewable Feedstock

Hrudayanath Thatoi and Sandesh Behera
Maharaja Sriram Chandra Bhanja Deo University Baripada, Odisha, India

3.1 INTRODUCTION

Cellulose is a major biopolymer due to its significant potential for various applications. It is the most prevalent substance on earth, and is produced in plants during photosynthesis. It is a linear homopolysaccharide, with β-D-glucopyranose units that are connected by β-1,4-glycosidic bonds. Bacteria and green plants both consist of cellulose, which is the most significant structural element (Kalia et al. 2011). Chawla et al. (2009) reported that various Gram-negative bacteria, including *Acetobacter, Aerobacter, Achromobacter, Agrobacterium, Azotobacter, Rhizobium, Pseudomonas, Sarcina*, and *Salmonella*, are capable of producing cellulose. Since it is produced by bacteria, the term given to the substance is bacterial cellulose (BC). In contrast to plant cellulose, BC is secreted without any byproducts such as lignin and hemicellulose and thus has a high purity. Therefore, BC requires no further purification process to remove the unwanted hemicellulose and lignin components as in the case of plant cellulose (Barhoum et al. 2020).

Being pure cellulose, BC is utilized to produce high-end cellulose-based products. It has the same molecular structure as natural cellulose from plants, but differs in its chemical and physical characteristics. BC, being free from wax, lignin, and hemicellulose, has a high purity that has several industrial applications. It is made up of nanofibrils that range in size from 2–4 nm that are bundled into nanoribbons of 60 nm in diameter (Cannon and Anderson 1991; Yamanaka et al. 1989). The unique nanostructure of BC shows high levels of polymerization, crystallinity, water holding capacity, and mechanical strength because of the glucose monomer in its structure. Because of its reticulated networks of thin fibers, which have a diameter of about 0.1 μm, 100 times smaller than plant-derived fiber, BC displays better tensile strength and water holding capacity. This is why many industries use bacterial cellulose as a raw material. Its strong tensile and water holding properties make it a crucial raw material for products including desserts, speakers, and high-acoustic papers (Gregory et al. 2021).

BC is generally nontoxic or exhibits low toxicity (Fu et al. 2013). It can be altered to have chemical and physical qualities that are suitable for reconstructive surgery (Kowalska-Ludwicka et al. 2013). It has been demonstrated that smooth muscle cells, chondrocytes, and endothelial cells attach to BC, thus it has very good biocompatibility (Zaborowska et al. 2010). BC is biocompatible and research has shown that it does not trigger an immune reaction other than temporary inflammation (Torres et al. 2012). Additionally, BC may be broken down by enzymes found in nature but not in humans (Fu et al. 2013). Much research is being conducted in an effort to enhance the physical and chemical characteristics of BC by combining it with natural sources such as protein, starch, betel leaves, gelatin, chitosan, and aloe vera, to enhance its potential for commercial applications. Research by Wiegand et al. (2006) showed that BC mixed with collagen significantly increased the antioxidant capabilities of BC film. Saibuatong and Phisalaphong (2010) created a nanostructure film using aloe vera and BC that had greater tensile strength and smaller pores compared to the unmodified BC film. Impregnation of BC with silver nanoparticles by Maneerung et al. (2008) was shown to enhance BC's antibacterial properties against both Gram-negative and Gram-positive

 DOI: 10.1201/9781003355434-3

strains of *Escherichia coli* and *Staphylococcus aureus*. Because of this discovery, the pharmaceutical and medical industries are showing interest in the uses of BC in the areas of health care. BC has been explored for use as a wound dressing in replacing vascular tubes and artificial skin. Researchers believe that BC may be used to treat surgical wounds, ulcers, bedsores, and to create new tissues and organs (Portela et al. 2019). Based on the widespread use of BC in several industrial sectors, researchers believe that more investigation into the manufacturing of BC would benefit society considerably. The present chapter provides an overview of BC structure, biosynthesis, production, and fermentation process. In addition, characteristics for different applications along with future prospects are also discussed.

3.2 STRUCTURE OF BACTERIAL CELLULOSE

Similar to plant cellulose, BC is a linear homopolymer with the molecular formula $(C_6H_{10}O_5)n$ and is made up of repeated units of glucose rings connected by β-1,4-glycosidic linkages. The molecular structure of BC comprises repeating β-D-glucopyaranosyl building blocks covalently connected with β-1,4-glycosidic bonds between the hydroxyl group of the C_4 and C_1 atoms. Several glucan monomers are held together by inter- and intra-hydrogen bonding that form a highly compact, fully extended cellulose sheet structure. These cellulose sheet structures are assembled by Van der Waals interaction and provide the crystalline structure of BC (Ul-Islam et al. 2012). BC differs from plant cellulose in its macromolecular structure and characteristics. Each BC chain is composed of one reducing end with a C_1-OH group and at the other end a C_4-OH group. The three-dimensional nanofibers that are aligned to form the fibrous network of BC lead to the formation of a hydrogel sheet with porosity and high surface area. During synthesis, protofibrils are released from glucose chains through the bacterial cell wall where they assemble to produce nanofibrils of cellulose ribbons (Ross et al. 1991). With a very porous matrix, these ribbons provide the web-shaped network structure of BC (Maria et al. 2010). In addition to varying with culture conditions, the thickness of the cellulose fiber varies from species to species. For example, the microfibril ribbons that make up BC fibers of *Gluconacetobacter xylinum* are 70–80 nm broad and 3–4 nm thick. BC microfibrils, which are about 100 times smaller than those of plant cellulose, were first described by Gayathry and Gopalaswamy (2014) and Chawla et al. (2009). Several hydroxyl groups are present on the surface of the cellulose, which provide the hydrophilic, biodegradable, and chemically modifiable nature to the cellulose (Klemm et al. 2005). The arrangement of the fibers plays an important role in the development of BC-based composites with various types of materials for different biomedical and optoelectronic applications (Ul-Islam et al. 2015). BC has a firm appearance due to the evenly dispersed and reticulate web-shaped structure of its fibers. The increased crystallinity, improved mechanical characteristics, and good thermal properties of BC are attributed to this fiber arrangement. Compared to plant cellulose, the small and thin ultrafine fibrous structure of BC gives it distinct features that provide the foundation for its wide range of applications.

3.3 BIOSYNTHESIS OF BACTERIAL CELLULOSE

In the biosynthesis of BC, different enzymes, as well as complexes of catalytic and regulatory proteins, are involved. During the biosynthesis, first uridinediphosphoglucose (UDP-glucose), the cellulose precursor, is produced, followed by the polymerization of glucose into the 1,4-glucan chain and nascent chain association into the distinctive ribbon-like structure, which is made up of approximately 100–1000 individual cellulose chains. Converting glucose to UDP-glucose via glucose-6-phosphate and glucose-1-phosphate, as well as cellulose synthase linking UDP-glucose to the end of a growing polymer chain is the basic enzyme-mediated process in the production of BC. Among the different enzymes, cellulose synthase (UDP-glucose: 1,4-β-D-glycosyltransferase; EC 2.4.1.12) is the enzyme considered to be most important to the production of BC. This enzyme is subjected to a complex regulatory system that manages and controls the enzyme's activation and inactivation (Römling and Galperin 2015).

A study of cellulose production by *Komagataeibacter xylinus* has resulted in the purification of the relevant enzyme complex and identification of its activator, cyclic dimeric (3′ to 5′) guanosine monophosphate (c-di-GMP). The pathways of c-di-GMP synthesis and degradation are catalyzed by enzymes that are present in cell-free extracts in both soluble and membrane-associated forms. A most soluble enzyme, diguanylatecyclase, helps in the condensation of two molecules of GTP in a two-step process using the intermediate linear triphosphate pppGpG, which is responsible for forming the nucleotide activator. A membrane-bound enzyme named PDE-A (a phosphodiesterase), degrades the activator and cleaves a single phosphodiester bond in the cyclic structure to form the linear 50-phosphoryl dimer pGpG (Ross et al. 1987).

The first breakdown product is further hydrolyzed to generate two molecules of 50-GMP in a mechanism linked to a second phosphodiesterase known as PDE-B, which has no stimulatory actions. Considering that the ratio of PDE-A to PDE-B activity in membrane preparations is around 10:1, the phosphodiesterase has been differentiated from PDE-A and PDE-B. Additionally, PDE-A activity is reduced at low Ca^{2+} ion concentrations, in contrast to the PDE-B response, indicating that this phosphodiesterase serves as an additional locus of regulatory control which is further maintained by Ca^{2+} ions.

Komagataeibacter xylinus previously referred to as *Gluconobacter/Acetobacter xylinus* is most known among the bacteria suitable for cellulose production. Biosynthesis of BC is a multistep pathway that is regulated by large numbers of singular enzymes and sets of regulatory proteins (Bielecki et al. 2005). *K. xylinus* is a model organism for studying bacterial cellulose synthesis because it contains the enzymes, genes, and biosynthesis mechanisms required for the process (Figure 3.1).

Generally, BC biosynthesis is being utilized as a simple model pathway to analyze the genetic aspects of other related biosynthesis cascades in plants that play significant roles in other plant metabolic pathways. Studies on BC biosynthesis have shed light on the causes of different environmental issues, such as disturbance in polysaccharide transport through the bacterial cell, which further result in loss of control in bacterial cell responses to oxygen and nitric oxide (NO). The loss of bacterial cell responses has been shown to be responsible for phenomena such as defects in cell

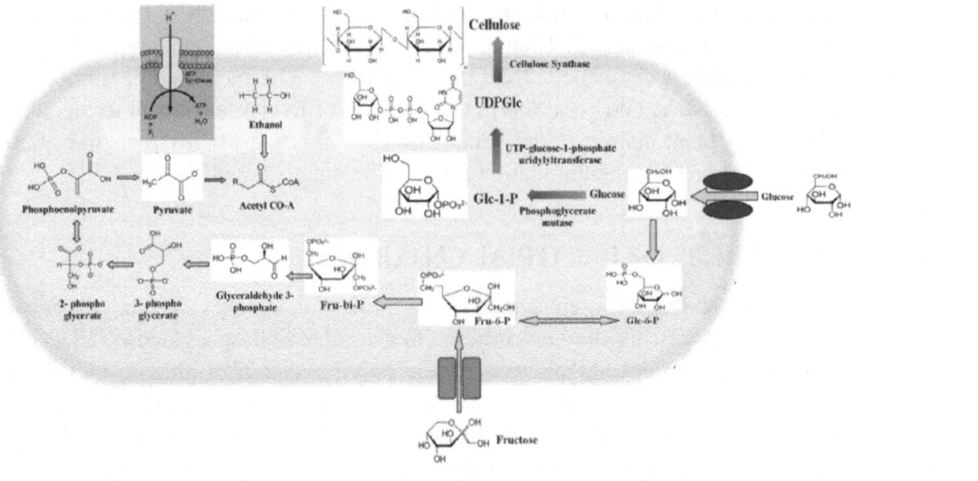

FIGURE 3.1 Bacterial cellulose production in *K. xylinus* involves the biosynthesis pathway and several different macromolecules.

Source: Adapted from Mishra et al. (2022.)

motility, cell-cell interactions, and biofilm formation and many other cellular stresses. Significant amounts of cellulose and its derivatives make up the extracellular matrix of biofilms, which serve crucial roles in limiting major plant and human diseases (Silva et al. 2016). Different methods have been employed in the research of BC synthesis; however, reconstructing whole metabolic pathways is now routinely performed by comparing the genome sequence to its biochemical and physiological data. Although substantial efforts have been made to improve BC in vitro production, it is still difficult to comprehend the system-level cellular metabolism involved in BC biosynthesis in the postgenomic age (Guerriero et al. 2016; Ullah et al. 2016).

3.4 PRODUCTION OF BACTERIAL CELLULOSE

BC production is a complicated process, influenced by many different factors and requires ideal circumstances to achieve maximal output. BC production depends on a number of variables that affect the productivity during the fermentation process, including the composition of the fermentation, the sources used in the medium, and the operating conditions, that is, temperature, pH, inoculation ratio, dissolved oxygen of the medium, and inoculum age. By maximizing these variables, bacterial cellulose production may be significantly increased. Bacterial cellulose manufacturing was started as early as 1886; thereafter, researchers from all over the world started working to improve the fermentation process in a variety of ways, including the culture method, medium components, and other elements that can increase BC synthesis. Industries also took interest in the research, because BC has several significant applications, including tissue engineering, pulp and paper manufacturing, biotechnology, medicine, and biopolymers (Chawla et al. 2009; Panesar et al. 2009; Bielecki et al. 2005; Keshk and Sameshima 2006).

There are several strains of bacteria from various genera that may make cellulose. But the bacteria belonging to the genus *Gluconacetobacter* (renamed *Komagataeibacter*) have been shown to produce significant amounts of cellulose. *Komagataeibacter* is frequently the genus of choice for study due to its higher BC productivity and purity as compared to other genera (Ruka et al. 2012). To date, many different strains of the genus *Komagataeibacter* have been discovered to produce membranes with varying network structures and mechanical characteristics (Chen et al. 2018). Other bacterial species that produce BC include *Achromobacter, Aerobacter, Agrobacterium, Azotobacter, Enterobacter, Rhizobium, Salmonella,* and *Sarcina. Gluconacetobacter xylinus* naturally creates cellulose biofilms on the surfaces of fruits and flowers. It is interesting to note that *G. xylinus* strains, which are well known for producing bacterial cellulose effectively, may also generate glucosidase.

Research into the use of affordable feedstock as a substitute for synthetic media is crucial with regard to these promising strains. Investigations have revealed that agricultural residues are excellent substrates for several biological activities because they are rich in lignocellulosic components

TABLE 3.1
Utilization of Different Feedstock for Bacterial Cellulose Production

Agro Wastes Utilized as Feedstocks	Microbe Involved in the Production of Cellulose	Maximum Productivity of Bacterial Cellulose (BC)
Miscanthus sacchariflorus	*Medusomyces gisevii* Sa-12	1.24 g/L
Sweet sorghum	*Acetobacter xylinum* ATCC 23767	-
Oil palm frond juice	*Acetobacter xylinum*	4.2 h/L
Oat hulls	*Medusomyces gisevii* Sa-12	2.20 g/L
Soybean oil refinery effluent	*Gluconacetobacter xylinus*	2.20 g/L

(*Continued*)

TABLE 3.1 (*Continued*)
Utilization of Different Feedstock for Bacterial Cellulose Production

Agro Wastes Utilized as Feedstocks	Microbe Involved in the Production of Cellulose	Maximum Productivity of Bacterial Cellulose (BC)
Apple pomace	*Gluconacetobacter xylinus*	7.19 g/L
Miscanthus straw, oat hulls	*Medusomyces gisevii* Sa-12	-
Oat hulls	*Medusomyces gisevii* Sa-12	2.2 g/L
Tobacco waste	*Acetobacter xylinum* ATCC 23767	5.2 g/L
Orange peel	*Gluconacetobacter xylinus*	6.13 g/L
Pecan nutshell	*Gluconacetobacter entanii*	2.81 g/L
Potato peel	*Gluconacetobacter xylinum* ATCC 10245	2.61 g/L
Lemon, mandarin, orange, grapefruit	*Komagataeibacter hansenii* GA2016	3.92 BC/100 g peel
Rotten banana/rotten mango	*Komagataeibacter medellinensis*	4.81 g/L
Sugarcane molasses	*Komagataeibacter rhaeticus*	1.90 g/L
Cheese, curd whey	*Gluconacetobacter sucrofermentans* B11267	5.45 g/L
Ripe dates	*Acetobacter xylinum* 0416	5.8 g/L
Durian shell	*Gluconacetobacter xylinus* CH001	2.67 g/L
Cashew tree residues	*Komagataeibacter rhaeticus*	6.0 g/L
Pineapple waste, pawpaw, watermelon	*Acetobacter pasteurianus* PW1	3.9 g/L
Orange peel	*Gluconoacetobacter xylinus* BCRC 12334	3.40 g/L
Corn stalk	*Acetobacter xylinum* ATCC 23767	2.86 g/L
Bagasse	*Gluconacetobacter xylinus* CH001	1.09 g/L
Apple pomace	*Gluconacetobacter medellinensis*	2.5 g/L
Grape pomace	*Acetobacter xylinum NRRL*	6.56 g/L
Citrus peel, pomace	*Komagataeibacter xylinus* CICC No. 10529	5.7 g/L
Pineapple waste	*Gluconacetobacter medellinensis*	4 g/L
Sugarcane juice	*Gluconacetobacter medellinensis*	0.8 g/L
Candied jujube processing wastewater	*Gluconacetobacter xylinum* CGMCC No. 2955	2.25 g/L
Sugarcane juice, pineapple residues	*Gluconacetobacter Medellinensis*	3.24 g/L
Pineapple, watermelon peels	*Komagataeibacter hansenii* MCM B-967	1.25 g/L
Pineapple peel	*Gluconacetobacter medellinensis*	3.97 g/L
Wheat straw	*Acetobacter xylinus* ATCC 23770	8.3 g/L
Coffee cherry husk	*Gluconacetobacter hansenii* UAC09	8.2 g/L
Elephant grass	*Gluconacetobacter xylinus* CH001	6.4 g/L
Spruce	*Gluconacetobacter xylinus* ATCC 23770	8.2 g/L
Grape bagasse	*Gluconacetobacter xylinus* NRRL B-42	8 g/L
Wheat straw	*Acetobacter xylinus* ATCC 23770	8.3 g/L
Coffee husk	*Gluconacetobacter hansenii* UAC09	5.6 g/L
Cotton waste textiles	*Acetobacter xylinus* ATCC 23770	10.8 g/L
Grape skins, cheese whey, crude glycerol	*Gluconacetobacter sacchari*	0.1 g/L
Pineapple peel, sugarcane juice	*Gluconacetobacter swingsii*	2.8 g/L
Maple syrup	*Acetobacter xylinum* BPR 2001	1.5 g/L
Grape skins	*Gluconacetobacter sacchari*	0.6 g/L
Orange, apple, pineapple Japanese pear, grape	*Acetobacter xylinum* NBRC 13693	5.9 g/L
Rice bark	*Acetobacter xylinum*	2.42 g/L
Fruit juices	*Acetobacter xylinum* NBRC 13693	5.9 g/L
Rice bark	*Acetobacter xylinum* ATCC 23769	2.42 g/L
Konjac powder	*Acetobacter aceti* ATCC 23770	2.12 g/L
Corn steep liquor	*Gluconacetobacter xylinum* BPR2001	8 g/L

(Table 3.1). Agricultural wastes and industrial byproducts such as rice husk, wheat straw, fruit peels and juice, palm date fruits, molasses, and olive oil mill effluent, which are renewable and inexpensive, have been investigated as BC production substrates in the past. As there is growing interest in the economical uses of BC, new cost-effective production strategies employing novel and affordable feedstock and the scaling up of bioprocess techniques for industrial applications have become increasingly important.

A naturally occurring nanomaterial (BC) is made by the members of the genus *Komagataeibacter* when they are grown in a medium containing carbon and a nitrogen source (Rajwade et al. 2015). For the past 10 years, significant global environmental, economic, and energy issues have increased the importance of long-term feedstock management and researchers are focusing on the use of renewable natural resources to make BC, thanks to advancements in biotechnological processes. The utilization of these natural wastes in BC manufacturing will reduce waste disposal costs, along with helping to clean the environment. For growth and cellulose generation, the *Enterobacter amnigenus* GH-1 strain was exposed to a variety of natural carbon sources, including starch hydrolysate, molasses, coconut milk, coconut water, sugarcane juice, orange juice, pineapple juice, and pomegranate juice. *Gluconacetobacter swingsii* developed bacterial cellulose microfibrils from unusual sources such agroindustrial remnants of sugarcane juice and pineapple peel. Additionally, batch fermentations using the bacterial strain *Komagataeibacter sucrofermentans* using commercial sugars and crude glycerol have been shown to produce the extracellular polysaccharide. The BC made from the Egyptian strain of *Achromobacter* sp. was free of impurities and had a pure structure.

Alternative low-cost substrates for BC manufacture have included certain inexpensive agricultural and industrial wastes and materials. For the first time, BC was produced using processed agricultural straw as a possible cost-effective feedstock. Additionally, BC was produced using a static cultivation method, with the byproduct of beer culture broth as the medium. Andritsou et al. (2018) demonstrated exceptional performance when they employed orange peel waste as a substrate for the manufacture of BC, with yields that were up to six times greater than those from traditional media. Citrus trash is a strong contender for use as a medium for BC synthesis due to its vast availability, low cost, renewability, and high carbohydrate content. To produce BC in a sustainable manner, feedstock from agroindustrial waste and byproduct industries must be used. For example, BC (around 13 g/L) has been produced in static cultures using crude glycerol from biodiesel production processes and waste streams from the flour-rich confectionery sector.

BC is also produced by acetic acid bacteria in oxidative fermentation using both synthetic and nonsynthetic culture media (Song and Kim 2019). *Acetobacter xylinum* can digest a wide range of sugars and create significant amounts of cellulose in liquid media at pH 3.0–7.0 and temperatures between 25°C and 30°C. The aerobic Gram-negative bacteria were found to actively ferment saccharides as a carbon source (Ross et al. 1991; Sani and Dahman 2010). In another technique, tea fungus, a symbiotic culture of acetic acid, bacteria, and yeast, is utilized to produce BC. Tea fungus has an extraordinary capacity to create a floating cellulose network over an unagitated substrate that mimics a surface mold, hence it is used to cultivate bacteria (Blanc 1996).

BC has also been produced industrially and is utilized extensively in several areas. According to a report by ResearchMoz, BC products have seen tremendous global success, particularly in the food sector. In 2016, the BC market had a value of about US $207.36 million. That value was predicted to rise to $497.76 million in 2022 and $700 million, or more, by 2026. Dourado et al. (2016) used Super-Pro Designer software to conduct a techno-economic analysis of the industrial-scale fermentation of BC. The software calculated that industrial companies need to invest around US $13 million to produce 504 tons of BC annually. The cost of annual production of BC is estimated to amount to $7.4 million and its annual net profit to be $3.3 million. Although BC production is capital intensive, manufacturers and researchers are attempting to find innovative ways to cut costs by isolating strains with high yields, optimizing fermentation reactors, and using inexpensive nutritional substrates.

Acetobacter sp. was used in the fermentation process that produced the first bacterial cellulose with the addition of oxygen and glucose (Chawla et al. 2009). Since then, a number of potential methods for mass producing BC have been discovered for use in a variety of industrial settings. Parameters such as temperature, pH of the medium, nitrogen source concentration, and concentration of carbon sources all have an impact on the amount of BC produced through the fermentation process (Jagannath et al. 2008). Also, the culture method also had an impact on the quantity and quality of BC produced during the fermentation process. Researchers have made several attempts to investigate BC synthesis utilizing different bacteria strains. Gram-negative bacteria may manufacture cellulose, including *Agrobacterium*, *Achromobacter*, *Aerobacter*, *Sarcina*, *Azotobacter*, *Rhizobium*, *Pseudomonas*, and *Salmonella* (Saibuatong and Phisalaphong 2010). *Sarcinaventriculi*, a Gram-positive bacterium, can also manufacture cellulose. According to reports, *Acetobacter xylinum*, *Acetobacter hansenii*, and *Acetobacter pasteurianus* are the most productive producers of bacterial cellulose (Chawla et al. 2009). Compared to other bacteria, *Acetobacter* is widely recognized for making bacterial cellulose with greater yields. *Acetobacter,* also known as *Gluconacetobacter xylinus*, is a rod-shaped, Gram-negative bacterium that can grow on a broad range of substrates and is able to produce BC while having no cellulase activity (Keshk and Sameshima 2006; Ross et al. 1991). BC has been synthesized from different types of carbon sources, including glucose, sucrose, fructose, and others. *Acetobacter* was shown to have 50% efficiency in the conversion of glycerol, hexoses, pyruvate, dihydroxyacetone, and dicarboxylic acids into BC. Additionally, researchers regard *Acetobacter* as the best candidate for the large-scale production of BC. Therefore, being pertinent to and appropriate for producing BC, *Acetobacter xylinum* was chosen as the bacterial strain for this purpose. According to Donini et al. (2010), when the production of BC reaches up to 15 g/L in 50 hours, the fermentative production of BC can achieve a production efficiency equivalent to the growth of plant cellulose. Additionally, the production area for BC fermentation is significantly lower than the area required for plant growth, and the process of purifying BC is easier and less polluting than the process of extracting cellulose from wood.

According to Chawla et al. (2009) in their analysis of bacterial cellulose fermentation processes, the medium used in the fermentation process is one of the primary sources of nutrients required by *Acetobacter* strains to synthesize cellulose throughout the culture time. Basically, potassium, magnesium salts, carbon, nitrogen, phosphorus, and sulfur are the nutrients needed for the process. To encourage cell development and improve the generation of bacterial cellulose, additional vitamins and amino acid supplements are also employed. Hestrin-Schramm (HS) medium, which was first established in 1954, is the standard medium used for bacterial cellulose synthesis. The HS medium is composed of glucose (20 g/L) as the source of carbon, yeast extracts (5 g/L), and peptone (5 g/L) as nitrogen sources, and citric acid (2.7 g/L), disodium phosphate (1.17 g/L) as other accessory components (Panesar et al. 2009). Using *Acetobacter* strains, which were utilized often in earlier experiments, the medium proved successful in providing the nutrients for BC production. The cost of the HS medium's components, however, makes it difficult for industry to produce vast amounts of BC on a large scale. Numerous forms of byproduct media, which primarily originate from agricultural waste, were researched in an effort to lower the cost of the medium. According to findings from earlier research, agricultural wastes such as maize steep liquor, sugarcane molasses, beet molasses, pineapple, coconut water, and empty palm oil fiber might stimulate the growth of *Acetobacter* strains and make it considerably capable of producing BC).

It has been found that the agricultural waste needed to be saccharified before being employed in medium production. In some situations, such as when utilizing palm oil fiber, many treatments must be carried out to obtain the proper conditions for medium production. As a result, preparing media becomes more difficult, time consuming, and expensive. Because the response surface methodology (RSM) requires running numerous repetitions of a fermentation process, the notion of employing widely available media is the ideal alternative for this investigation. Coconut water is one of the simple-to-make substances discussed before. Since a very long time ago, coconut water has been utilized as the manufacturing medium for vinegar and nata de coco. The efficacy of coconut water to produce nata de coco without the use of additional nutrient supplements or carbon sources has been demonstrated by the fact that rural residents of India and the Philippines were able to use coconut water as a medium to ferment BC into a gelatinous form. Additionally, prior investigations into BC fermentation utilizing coconut water as the medium were carried out sequentially and yielded a sizable quantity of bacterial cellulose (Saibuatong and Phisalaphong 2010).

In addition to the fermentation parameters, such as carbon source concentration, temperature, and pH, the culture method also influences the quality and volume of BC production. According to the findings, agitated and reactor cultures generated more bacterial cellulose than static cultures. However, static cultures produce bacterial cellulose of a far higher caliber than alternative methods. Bacterial cellulose synthesized in static culture has demonstrated the manufacture of rigorous fibrils with enormous cell size. Bacterial cellulose created by static culture also has a better capacity to retain water than bacterial cellulose created through other cultivation methods.

3.5 PATENTS ON BACTERIAL CELLULOSE

The influence of the expanding worldwide market resulted in the filing of more than 6301 patents or more related to BC between 2009 and 2019. China, the United States, and Japan are the nations with the most patents that have been registered. These patents range from novel techniques of producing BC to new products such as anti-inflammatory and antiallergenic face masks, BC-based automotive coatings, and BC-chitosan dressings (Fernandes et al. 2020).

Each year, more and more patents relating to nanocellulose are published. The discovery of new patents on the development of nanocellulose is now trending worldwide, and involves cellulose nanocrystals and microfibrillated and bacterial cellulose. Between 2019 and 2022, there were ten patents for inventions for both products and methods (Table 3.2). In 2022, a patent was filed on the development of nanocellulose-reinforced corrugated medium. In the same year, another patent application was filed for a novel technique based on the use of bacterial cellulose hydrogel to produce a biocomposite with antibacterial characteristics.

3.6 APPLICATIONS OF BACTERIAL CELLULOSE

BC is beneficial in a wide range of applications due to its adaptability and different characteristics, including porosity, high degree of purity, high water retention capacity, ultrafine work, and long-term stability (Ullah et al. 2016). Due to BC's identical chemical structure to plant cellulose but superior in characteristics, it is more helpful in a variety of applications, including drug delivery systems, medicine, food, and cosmetics (Lin et al. 2013). In the past several decades, the widespread use of cellulose in biomedical applications has been encouraged due to its biocompatibility, biodegradability, and cost effectiveness. It exhibits the characteristics of a suitable scaffold to promote the development of cells and tissues. Table 3.3 provides more details on the many uses of BC.

3.6.1 Biomedical Fields

BC has received a lot of attention for its potential in medical applications (Kucińska-Lipka et al. 2015). Currently, due to BC's great biocompatibility and capacity to provide the ideal three-dimensional substrate for cell attachment, wound dressing and tissue engineering are now its primary commercial applications (Picheth et al. 2017). In the pharma industry, BC is suggested as support for transdermal drug delivery (Mohd et al. 2014). A variety of trademarks, including XCell, Bioprocess, Gengiflex, Biofill, Membracell, and Bionext, are used to market BC as wound dressing devices. These products have the benefit of adapting remarkably well to nearly any contour since they can maintain an ideal water balance by either absorbing or releasing fluid in accordance with the behavior of the wound. Tissue engineering is another innovative sector with fascinating prospective uses for BC material because of its remarkably low cytotoxicity, great mechanical

TABLE 3.2
Recent Bacterial Cellulose Patents

Publication Number	Reference Title	Assignee	Inventors	Year of Publication
US11359334B2	Nanocellulose-reinforced corrugated medium	GranBio Intellectual Property Holdings, LLC, Thomaston, GA (US)	Theodora Retsina, Atlanta, GA (US); (56) References Cited Kimberly Nelson, Atlanta, GA (US); Lee Hill, Atlanta, GA (US); Vesa Pylkkanen, Atlanta, GA (US); Timothy J. Galliford, Newnan, GA (US)	2022
RU2771864C1	Method for obtaining a biocomposite with antibacterial properties based on bacterial cellulose hydrogel	Federalnoegosudarstvennoebiudzhetnoe, obrazovatelnoeuchrezhdenievysshego, obrazovaniia "Natsionalnyiissledovatelskii, Mordovskiigosudarstvennyiuniversitetim. N.P., Ogareva" (RU)	Revin Viktor Vasilevich (RU), Liiaskina Elena Vladimirovna (RU), Khramova Oksana Olegovna (RU)	2022
US2022/0034038A1	Sulfite-based processes for producing nanocellulose, and compositions and products produced therefrom	GranBio Intellectual Property Holdings, LLC, Minnetrista, MN (US)	Kimberly Nelson, Atlanta, GA (US); Theodora Retsina, Atlanta, GA (US); Vesa Pylkkanen, Atlanta, GA (US); Ryan O'Connor, Minnetrista, MN (US)	2022
US2021/0363558 A1	Compositions for saccharification of cellulosic material	Novozymes, Inc., Davis, CA (US)	Brett McBrayer, Davis, CA (US); Tarana Shaghasi, Davis, CA (US); Elena Vlasenko, Davis, CA (US)	2021
US2021/0284757 A1	Processes and apparatus for producing nanocellulose, and compositions and products produced therefrom	GranBio Intellectual Property Holdings, LLC, Minnetrista, MN (US)	Kimberly Nelson, Atlanta, GA (US); Theodora Retsina, Atlanta, GA (US); Vesa Pylkkanen, Atlanta, GA (US); Ryan O'Connor, Minnetrista, MN (US)	2021
US2021/0254285 A1	Cellulose materials and methods of making and using the same	Ecoinno (H.K) Limited, Hong Kong (CN)	George Chen, Hong Kong (CN); Koon- Yang Lee, London (GB); Alexander Bismarck, Vienna (AT); Robert Li, Hong Kong (CN)	2021
US10603651B2	Process and system for producing pulp, energy, and bioderivatives from plant-based and recycled materials	Tyton Biosciences, LLC, Danville, VA (US)	Sandeep Kumar, Virginia Beach, VA (US); Peter Majeranowski, Danville, VA (US); Igor Kostenyuk, Winter Haven, FL (US); Iulian Bobe, Danville, VA (US); Florin Barla, Danville, VA (US)	2020

WO2020/136629A1	Bacterial cellulose formulations, methods and uses thereof	Universidase Do Mintho, Largo Do Paco, 4704–553 Braga (PT)	Queiros Dourado, Fernando Octavio,	2020
US2019/0055373 A1	Bacterial cellulose gels, process for producing and methods of use	The Regents of the University of Colorado, a Body Corporate, Denver, CO (US)	Andrew J. Hess, Louisville, CO (US); Ivan I. Smalyukh, Boulder, CO (US); Qingkun Liu, Boulder, CO (US); Joshua A. De La Cruz, Denver, CO (US); Blaise Fleury, Boulder, CO (US); Eldho Abraham, Boulder, CO (US); Vladyslav Cherpak, Westminster, CO (US); Bohdan Senyuk, Boulder, CO (US)	2019
EP3572043A1	Medical implant with seamlessly connected bacterial cellulose	Galander, Marcus Biotronik Corporate Services SE Corporate Intellectual Property Sieversufer 7–912359 Berlin (DE)	Alexander Rzany, Tina Pieger, Bernhard Hensel	2019

TABLE 3.3
Industrial Applications of Uses of Bacterial Cellulose

Area	Application	References
Biomedical	• Wound dressing	(Fürsatz et al. 2018)
	• Burn treatment	(Silveira et al. 2016)
	• Skin repair	(Lina et al. 2011)
	• Biosensing film/biosensor	(Liu et al. 2017, 2018)
	• Living wearable device	
	• Cell encapsulation	
Tissue engineering	• Temporary artificial skin	(Wu et al. 2018)
	• Dental implant components	(Weyell et al. 2019)
	• Artificial blood vessels BASYC	(Klemm et al. 2005)
	• Contact lenses	(Levinson and Glonek 2010)
Pharmaceutical industry	• Drug delivery	(Bacakova et al. 2019)
	• Drug excipient per oral (tablets)	
	• Paracetamol	
	• Ibuprofen (colonic release)	
	• Methotrexate (colonic release)	
Cosmetics	• Skin mask	(Pacheco et al. 2018)
	• Facial scrub	(Hasan et al. 2012)
Electronics	• Electronic paper display	(Aleshin et al. 2017)
	• Conductive membrane and material	(Zhou et al. 2013)
	• Battery separator	(Jang et al. 2017)
Textiles	• Clothing fabric	(Chan et al. 2018)
	• Coloring material	(Han et al. 2019)
Food industry	• Edible cellulose	(Shi et al. 2014)
	• Thickening agent	(Shi et al. 2014)
	• Gelling agent	(Ullah et al. 2016)
	• Stabilizing agent	(Gao et al. 2014)
	• Emulsifying and suspending agents	
	• Water-holding agent	
	• Food packaging incorporating antimicrobial agents	
Topical applications	• Local anesthetics	(Bacakova et al. 2019)
	• Antiseptics	
	• Antibiotics	
	• Antibacterial peptides	
	• Other antitumor, anti-inflammatory, and antimicrobial drugs	
Energy	• Microbial fuel cells/bioanodes	(Mashkour et al. 2016)
	• Proton-permeable membrane for Microbial Fuel Cells	(Ko et al. 2015)
	• Electrodes and supercapacitors	(Kim et al. 2019)
Carrier for immobilization	• Immobilization of enzymes	Wu et al. (2018)

capabilities, and highly porous structure (Torgbo and Sukyai 2018) and also showed its importance in the treatment of burns. According to Silveira et al. (2016), BC membranes speed the healing of second-degree burns. Various initiatives have been undertaken to increase the biological uses of BC, including (1) the production of functional polymers such as collagen for use in BC composites (Zhijiang and Guang 2011) and (2) the development of BC-based biomedical devices, biosensing film, and biosensors (Liu et al. 2017).

3.6.2 Drug Delivery

BC is considered to be a biodegradable, affordable, and biocompatible natural substance that has the potential to be used for drug delivery. One of the most significant manufacturers of Malaysia explored the concept of nata de coco (coconut gel). The capacity of BC to make soft, flexible, and foldable films was discovered to be exceptional; it was coupled with various amounts of acrylic acid to create hydrogels that were then exposed to increased electron beam irradiation at various dosages. Thermal and morphological analysis showed that thermally stable hydrogels were formed, with the amount of acrylic acid and the radiation dosage influencing the size of the pores. The hydrogels were shown to be thermo- and pH-responsive according to the results of studies on swelling and in vitro drug release (Amin et al. 2012). Kralisch at Friedrich-Schiller University in Jena, Germany, focuses on the same area and studies antiseptic small- and medium-sized molecules, healing proteins, antibiotics, and associated nanoparticles. According to the Ritger-Peppas equation, the release of the antiseptic medication (Octenidine) was based on diffusion and swelling. It showed a time-dependent biphasic release profile with a quick release in the first 8 hours and a delayed release rate up to 96 hours (Moritz et al. 2014).

3.6.3 Cell Encapsulation

The components of a tissue engineering construct include scaffolds, growth factors, and cells. In a conventional tissue engineering process, cells are seeded on the surface of a porous three-dimensional substrate with the appropriate shape and mechanical characteristics. In this situation, cell-cell and cell-matrix interactions mostly take place in two dimensions. However, under in vivo conditions, the extracellular matrix–embedded cells can generate a three-dimensional environment for these interactions. Researchers have started to encapsulate viable, functional cells in conditions that replicate the extracellular matrix found in nature. Using a variety of techniques, cells have been enclosed in three-dimensional structures, ranging from simple gravitational dripping to more advanced systems such as micro technologies or additive printing. For several applications, this method has shown interesting and promising outcomes, such as (1) cell immune isolation and drug delivery, (2) storage and transport, and (3) cell delivery. Natural hydrogels are desirable materials as they may be processed under mild conditions that are favorable to cells and have a structure that is comparable to that of the extracellular matrix. Additionally, the hydrogels are extremely biocompatible due to the low interfacial tension between them and the surrounding fluids, which results in a low rate of protein adsorption and cell adhesion. These flexible and malleable qualities of the gel lower the mechanical irritability on the surrounding tissue while providing a high degree of permeability to nutrients and metabolites.

3.6.4 Cosmetic Fields

BC is a great biomaterial for the cosmetics industry due to its superior water-holding capacity, nontoxicity, and lack of allergic side effects (Lin et al. 2013). Because of the membrane-like shape of BC, which gives it excellent mechanical properties, malleability, and ease of handling, the use of cellulose fibrils in cosmetics provides the benefit of a skin-care product that is nonirritating and takes the form of a stable oil-water emulsion without the need for additional surfactant (Amnuaikit et al. 2011). Innovative face masks have been made using BC as the support for the cosmetic actives (Muhsinin et al. 2017). Amnuaikit et al. (2011) investigated BC skin masks that provided more skin moisture absorption than moist towel application. In two cosmetic mask compositions, two active ingredients were integrated into BC membranes, according to Pacheco et al. (2018). The first involved a hydroviton-based skin-moisturizing association, while the second included propolis extract, an antibacterial and astringent active ingredient. Cosmetic companies BOTANISIGN, Bel Mondo Beauty LLC, Mary Kay, DHC, and STARSKIN have a substantial market for biocellulose masks in the commercial cosmetic sectors. To develop a novel, specially designed face scrub that comprises BC as its principal constituent, Hasan et al. (2012) explored BC as a natural facial scrub. This can be explained by the viscosity profile being comparable to that of commercial face scrubs (Keshk 2014; Kucińska-Lipka et al. 2015; Lina et al. 2011).

3.6.5 Electronic Materials and Devices

Due to its high reflectiveness, flexibility, nanofibrous porous network structure, light mass and ease of mobility, purity, and broad viewing angles, BC has found utility in electronics (Mohite and Patil 2014). Applications for electronic paper display are determined by these distinctive features of BC (Aleshin et al. 2017). Other uses are conductive membranes (Zhou et al. 2013), battery separators (Jang et al. 2017), and biosensors (Liu et al. 2018). By immobilizing electrochromic dyes within the microstructure and then depositing ions everywhere over the microfibrils to create a conducting channel, an electricity-conducting (or semiconducting) BC membrane was created. Multi-walled carbon nanotubes were also integrated by Yoon et al. (2006) into BC pellicles to create highly electrically conductive polymeric membranes. Numerous further uses for BC are possible, including e-book tablets, e-newspapers, dynamic wall murals, rewriteable maps, and educational tools. Olsson et al. (2010) created lightweight porous magnetic aerogels that can be compressed into stiff magnetic nanopaper using freeze-dried BC nanofibril aerogels as templates. Because of its high tensile strength, high thermostability, and high crystallinity, an example of an ideal use of BC is as a battery separator in electrochemical energy storage systems, such as Li^{3+} batteries (Keshk 2014). For the battery separator to transport ions effectively while preventing direct contact between the anode and cathode, it must be physically and chemically stable (Jang et al. 2017). The battery's performance is strongly influenced by its structure and qualities. BC is reported as an excellent contender for the next-generation battery separator due to its distinctive fibrous and cross-linked three-dimensional network structure (Jang et al. 2017).

3.7 CONCLUSIONS

Over the past few years, distinctive features of bacterial cellulose (BC) have received a lot of attention. Particularly appealing features that make bacterial nanocellulose an intriguing alternative for the environmentally friendly manufacturing of cellulose nanofibers are its design and high purity. BC will undoubtedly be a crucial material for future innovation and high-tech applications. In the biological, cosmetic, electrical, and fashion industries, BC has a lot of potential. With regard to all of the aforementioned categories, there are still certain obstacles to be solved, such as porosity control, regulated biodegradation, and low productivity rate of BC. Shorter incubation times, reduced production costs, and more cost-effective large-scale production conditions must be created for the critical variables. The majority of bacterial cellulose research publications published in the past 10 years have concentrated mostly on isolation, bioprocess development for its synthesis, and applications. It has been discovered that there is a high demand for bacterial cellulose because of its incredibly intriguing qualities, which have permitted its use in a wide range of applications. Economical production of BC would broaden its application as novel types of biomaterial. To make this biotechnological material commercially viable, economically practical, and a competitive product, additional research must be done to increase its capacity for production.

3.8 REFERENCES

Aleshin, A.N., Shcherbakov, I.P., Khripunov, A.K., et al., (2017). Light-emitting flexible transparent paper based on bacterial cellulose modified with semiconducting polymer MEH: PPV. Flexible and Printed Electronics, 2(3), p. 035004.

Amin, M.C.I.M., Ahmad, N., Halib, N., (2012). Synthesis and characterization of thermo-and pH-responsive bacterial cellulose/acrylic acid hydrogels for drug delivery. Carbohydrate Polymers, 88(2), pp. 465–473.

Amnuaikit, T., Chusuit, T., Raknam, P. (2011). Effects of a cellulose mask synthesized by a bacterium on facial skin characteristics and user satisfaction. Medical Devices (Auckland, NZ), 4, p. 77.

Andritsou, V., De Melo, E.M., Tsouko, E., et al., (2018). Synthesis and characterization of bacterial cellulose from citrus-based sustainable resources. ACS Omega, 3(8), pp. 10365–10373.

Article I. Dow, J.M., Fouhy, Y., Lucey, J., Ryan, R.P., (2009). Cyclic di-GMP as an intracellular signal regulating bacterial biofilm formation. Lab-on-a-Chip, 10(238).

Bacakova, L., Pajorova, J., Bacakova, M., et al., (2019). Versatile application of nanocellulose: From industry to skin tissue engineering and wound healing. Nanomaterials, 9(2), p. 164.

Barhoum, A., Jeevanandam, J., Rastogi, A., et al., (2020). Plant celluloses, hemicelluloses, lignins, and volatile oils for the synthesis of nanoparticles and nanostructured materials. Nanoscale, 12(45), pp. 22845–22890.

Bielecki, S., Krystynowicz, A., Turkiewicz, M., (2005). Bacterial Cellulose. Polysaccharides and Polyamides in the Food Industry: Properties, Production, and Patents, pp. 31–84.

Blanc, P.J., (1996). Characterization of the tea fungus metabolites. Biotechnology Letters, 18(2), pp. 139–142.

Cannon, R.E., Anderson, S.M., (1991). Biogenesis of bacterial cellulose. Critical Reviews in Microbiology, 17(6), pp. 435–447.

Chan, C.K., Shin, J., Jiang, S.X.K., (2018). Development of tailor-shaped bacterial cellulose textile cultivation techniques for zero-waste design. Clothing and Textiles Research Journal, 36(1), pp. 33–44.

Chawla, P.R., Bajaj, I.B., Survase, S.A., (2009). Microbial cellulose: Fermentative production and applications. Food Technology & Biotechnology, 47(2).

Chen, S.Q., Lopez-Sanchez, P., Wang, D., (2018). Mechanical properties of bacterial cellulose synthesised by diverse strains of the genus Komagataeibacter. Food Hydrocolloids, 81, pp. 87–95.

Donini, Í.A., De Salvi, D.T., Fukumoto, F.K., et al., (2010). Biosynthesis and recent advances in production of bacterial cellulose. Eclética Química, 35, pp. 165–178.

Dourado, F., Fontão, A., Leal, M., (2016). Process modeling and techno-economic evaluation of an industrial bacterial nanocellulose fermentation process. In Bacterial Nanocellulose (pp. 199–214). Elsevier.

Fernandes, I.D.A.A., Pedro, A.C., Ribeiro, V.R., et al., (2020). Bacterial cellulose: From production optimization to new applications. International Journal of Biological Macromolecules, 164, pp. 2598–2611.

Fu, L., Zhang, J. and Yang, G., (2013). Present status and applications of bacterial cellulose-based materials for skin tissue repair. Carbohydrate Polymers, 92(2), pp. 1432–1442.

Fürsatz, M., Skog, M., Sivlér, P., et al., (2018). Functionalization of bacterial cellulose wound dressings with the antimicrobial peptide ε-poly-L-Lysine. Biomedical Materials, 13(2), p. 025014.

Gao, C., Yan, T., Du, J., et al., (2014). Introduction of broad spectrum antibacterial properties to bacterial cellulose nanofibers via immobilising ε-polylysine nanocoatings. Food Hydrocolloids, 36, pp. 204–211.

Gayathry, G., Gopalaswamy, G., (2014). Production and characterisation of microbial cellulosic fibre from Acetobacter xylinum.

Gregory, D.A., Tripathi, L., Fricker, A.T., et al., (2021). Bacterial cellulose: A smart biomaterial with diverse applications. Materials Science and Engineering: R: Reports, 145, p. 100623.

Guerriero, G., Hausman, J.F., Strauss, J., (2016). Lignocellulosic biomass: Biosynthesis, degradation, and industrial utilization. Engineering in Life Sciences, 16(1), pp. 1–16.

Han, J., Shim, E., Kim, H.R., (2019). Effects of cultivation, washing, and bleaching conditions on bacterial cellulose fabric production. Textile Research Journal, 89(6), pp. 1094–1104.

Hasan, N., Biak, D.R.A., Kamarudin, S., (2012). Application of bacterial cellulose (BC) in natural facial scrub. International Journal on Advanced Science, Engineering and Information Technology, 2(4), pp. 1–4.

Jagannath, A., Kalaiselvan, A., Manjunatha, S.S., (2008). The effect of pH, sucrose and ammonium sulphate concentrations on the production of bacterial cellulose (Nata-de-coco) by *Acetobacter xylinum*. World Journal of Microbiology and Biotechnology, 24(11), pp. 2593–2599.

Jang, W.D., Hwang, J.H., Kim, H.U., (2017). Bacterial cellulose as an example product for sustainable production and consumption. Microbial Biotechnology, 10(5), p. 1181.

Kalia, S., Kaith, B.S., Kaur, I., eds., (2011). Cellulose Fibers: Bio-and Nano-Polymer Composites: Green Chemistry and Technology. Springer Science & Business Media.

Keshk, S.M., (2014). Bacterial cellulose production and its industrial applications. Journal Bioprocess Biotech, 4(2), p. 150.

Keshk, S.M., Sameshima, K., (2006). The utilization of sugar cane molasses with/without the presence of lignosulfonate for the production of bacterial cellulose. Applied Microbiology and Biotechnology, 72(2), pp. 291–296.

Kim, J.H., Lee, D., Lee, Y.H., (2019). Nanocellulose for energy storage systems: Beyond the limits of synthetic materials. Advanced Materials, 31(20), p. 1804826.

Klemm, D., Heublein, B., Fink, H.P., (2005). Cellulose: Fascinating biopolymer and sustainable raw material. Angewandte Chemie International Edition, 44(22), pp. 3358–3393.

Ko, Y.H., Oh, H.J., Lee, H.J., (2015). Use of bacterial cellulose from Gluconacetobacter hansenii NOK21 as a proton-permeable membrane in microbial fuel cells. Journal of Microbial Biochemistry and Technology, 7(3), pp. 145–151.

Kowalska-Ludwicka, K., Cala, J., Grobelski, B., et al., (2013). Special paper: New methods modified bacterial cellulose tubes for regeneration of damaged peripheral nerves. Archives of Medical Science, 9(3), pp. 527–534.

Kucińska-Lipka, J., Gubanska, I., Janik, H.J.P.B., (2015). Bacterial cellulose in the field of wound healing and regenerative medicine of skin: Recent trends and future prospectives. Polymer Bulletin, 72(9), pp. 2399–2419.

Levinson, D.J., Glonek, T., (2010). Microbial Cellulose Contact Lens. U.S. Patent 7,832,857.

Lin, W.C., Lien, C.C., Yeh, H.J., (2013). Bacterial cellulose and bacterial cellulose: Chitosan membranes for wound dressing applications. Carbohydrate Polymers, 94(1), pp. 603–611.

Lina, F., Yue, Z., Jin, Z., (2011). Bacterial cellulose for skin repair materials. In Biomedical Engineering: Frontiers and Challenges. InTech Publisher (pp. 249–274).

Liu, M., Liu, L., Jia, S., et al., (2018). Complete genome analysis of *Gluconacetobacter xylinus* CGMCC 2955 for elucidating bacterial cellulose biosynthesis and metabolic regulation. Scientific Reports, 8(1), pp. 1–10.

Liu, X., Tang, T.C., Tham, E., et al., (2017). Stretchable living materials and devices with hydrogel: Elastomer hybrids hosting programmed cells. Proceedings of the National Academy of Sciences, 114(9), pp. 2200–2205.

Maneerung, T., Tokura, S., Rujiravanit, R., et al., (2008). Impregnation of silver nanoparticles into bacterial cellulose for antimicrobial wound dressing. Carbohydrate Polymers, 72(1), pp. 43–51.

Maria, L., Santos, A.L., Oliveira, P.C., et al., (2010). Preparation and antibacterial activity of silver nanoparticles impregnated in bacterial cellulose. Polimeros, 20, pp. 72–77.

Mashkour, M., Rahimnejad, M., Mashkour, M., (2016). Bacterial cellulose-polyaniline nano-biocomposite: A porous media hydrogel bioanode enhancing the performance of microbial fuel cell. Journal of Power Sources, 325, pp. 322–328.

Mishra, S., Singh, P.K., Pattnaik, R., et al., (2022). Biochemistry, synthesis, and applications of bacterial cellulose: A review. Frontiers in Bioengineering and Biotechnology, 10.

Mohd Amin, M.C.I., Ahmad, N., Pandey, M., (2014). Stimuli-responsive bacterial cellulose-g-poly (acrylic acid-co-acrylamide) hydrogels for oral controlled release drug delivery. Drug Development and Industrial Pharmacy, 40(10), pp. 1340–1349.

Mohite, B.V., Patil, S.V., (2014). A novel biomaterial: Bacterial cellulose and its new era applications. Biotechnology and Applied Biochemistry, 61(2), pp. 101–110.

Moritz, S., Wiegand, C., Wesarg, F., et al., (2014). Active wound dressings based on bacterial nanocellulose as drug delivery system for octenidine. International Journal of Pharmaceutics, 471(1–2), pp. 45–55.

Muhsinin, S., Putri, N.T., Ziska, R., (2017). Bacterial cellulose from fermented banana peels (Musa paradisiaca) by Acetobacter xylinum as matrix of biocellulose mask. Journal of Pharmaceutical Sciences and Research, 9(2), p. 159.

Olsson, R.T., Azizi Samir, M.A.S., Salazar-Alvarez, G., et al., (2010). Making flexible magnetic aerogels and stiff magnetic nanopaper using cellulose nanofibrils as templates. Nature Nanotechnology, 5(8), pp. 584–588.

Pacheco, G., de Mello, C.V., Chiari-Andréo, B.G., et al., (2018). Bacterial cellulose skin masks: Properties and sensory tests. Journal of Cosmetic Dermatology, 17(5), pp. 840–847.
Panesar, P.S., Chavan, Y.V., Bera, M.B., (2009). Evaluation of Acetobacter strain for the production of microbial cellulose. Asian Journal of Chemistry, 21(10), pp. 99–102.
Picheth, G.F., Pirich, C.L., Sierakowski, M.R., et al., (2017). Bacterial cellulose in biomedical applications: A review. International Journal of Biological Macromolecules, 104, pp. 97–106.
Portela, R., Leal, C.R., Almeida, P.L., (2019). Bacterial cellulose: A versatile biopolymer for wound dressing applications. Microbial Biotechnology, 12(4), pp. 586–610.
Rajwade, J.M., Paknikar, K.M., Kumbhar, J.V., (2015). Applications of bacterial cellulose and its composites in biomedicine. Applied Microbiology and Biotechnology, 99(6), pp. 2491–2511.
Römling, U., Galperin, M.Y., (2015). Bacterial cellulose biosynthesis: Diversity of operons, subunits, products, and functions. Trends in Microbiology, 23(9), pp. 545–557.
Ross, P., Mayer, R., Benziman, M., (1991). Cellulose biosynthesis and function in bacteria. Microbiological Reviews, 55(1), pp. 35–58.
Ross, P., Weinhouse, H., Aloni, Y., et al., (1987). Regulation of cellulose synthesis in Acetobacter xylinum by cyclic diguanylic acid. Nature, 325(6101), pp. 279–281.
Ruka, D.R., Simon, G.P., Dean, K.M., (2012). Altering the growth conditions of Gluconacetobacter xylinus to maximize the yield of bacterial cellulose. Carbohydrate Polymers, 89(2), pp. 613–622.
Saibuatong, O.A., Phisalaphong, M., (2010). Novo aloe vera: Bacterial cellulose composite film from biosynthesis. Carbohydrate Polymers, 79(2), pp. 455–460.
Sani, A., Dahman, Y., (2010). Improvements in the production of bacterial synthesized biocellulose nanofibres using different culture methods. Journal of Chemical Technology & Biotechnology, 85(2), pp. 151–164.
Shi, Z., Zhang, Y., Phillips, G.O., (2014). Utilization of bacterial cellulose in food. Food Hydrocolloids, 35, pp. 539–545.
Silva, L.N., Zimmer, K.R., Macedo, A.J., (2016). Plant natural products targeting bacterial virulence factors. Chemical Reviews, 116(16), pp. 9162–9236.
Silva, S.S., Fernandes, E.M., Silva-Correia, J., et al., (2017). Natural-Origin Materials for Tissue Engineering and Regenerative Medicine. Universidade do Minho: RepositoriUM. doi: 10.1016/B978-0-12-803581-8.10134-1
Silveira, F.C.A., Pinto, F.C.M., Caldas Neto, S.D.S., et al., (2016). Treatment of tympanic membrane perforation using bacterial cellulose: A randomized controlled trial. Brazilian Journal of Otorhinolaryngology, 82, pp. 203–208.
Song, J.E., Kim, H.R., (2019). Bacterial cellulose as promising biomaterial and its application. In Advances in Textile Biotechnology (pp. 263–277). Woodhead Publishing.
Torgbo, S., Sukyai, P., (2018). Bacterial cellulose-based scaffold materials for bone tissue engineering. Applied Materials Today, 11, pp. 34–49.
Torres, F.G., Commeaux, S., Troncoso, O.P., (2012). Biocompatibility of bacterial cellulose based biomaterials. Journal of Functional Biomaterials, 3(4), pp. 864–878.
Ul-Islam, M., Khan, S., Ullah, M.W., (2015). Bacterial cellulose composites: Synthetic strategies and multiple applications in bio-medical and electro-conductive fields. Biotechnology Journal, 10(12), pp. 1847–1861.
Ul-Islam, M., Khan, T., Park, J.K., (2012). Water holding and release properties of bacterial cellulose obtained by in situ and ex situ modification. Carbohydrate Polymers, 88(2), pp. 596–603.
Ullah, H., Santos, H.A., Khan, T., (2016). Applications of bacterial cellulose in food, cosmetics and drug delivery. Cellulose, 23(4), pp. 2291–2314.
Weyell, P., Beekmann, U., Küpper, C., et al., (2019). Tailor-made material characteristics of bacterial cellulose for drug delivery applications in dentistry. Carbohydrate Polymers, 207, pp. 1–10.
Wiegand, C., Elsner, P., Hipler, U.C., (2006). Protease and ROS activities influenced by a composite of bacterial cellulose and collagen type I in vitro. Cellulose, 13(6), pp. 689–696.
Wu, C.N., Fuh, S.C., Lin, S.P., (2018). TEMPO-oxidized bacterial cellulose pellicle with silver nanoparticles for wound dressing. Biomacromolecules, 19(2), pp. 544–554.
Yamanaka, S., Watanabe, K., Kitamura, N., et al., (1989). The structure and mechanical properties of sheets prepared from bacterial cellulose. Journal of Materials Science, 24(9), pp. 3141–3145.
Yoon, S.H., Jin, H.J., Kook, M.C., (2006). Electrically conductive bacterial cellulose by incorporation of carbon nanotubes. Biomacromolecules, 7(4), pp. 1280–1284.

Zaborowska, M., Bodin, A., Bäckdahl, H., et al., (2010). Microporous bacterial cellulose as a potential scaffold for bone regeneration. Acta Biomaterialia, 6(7), pp. 2540–2547.
Zhijiang, C., Guang, Y., (2011). Bacterial cellulose/collagen composite: Characterization and first evaluation of cytocompatibility. Journal of Applied Polymer Science, 120(5), pp. 2938–2944.
Zhou, T., Chen, D., Jiu, J., et al., (2013). Electrically conductive bacterial cellulose composite membranes produced by the incorporation of graphite nanoplatelets in pristine bacterial cellulose membranes. Express Polymer Letters, 7(9).

4 Strategies for Scale-up of Bacterial Cellulose Production

Anu Radha
CSIR – Indian Institute of Integrative Medicine
Jammu, Jammu and Kashmir, India

Vivek Kumar Ahluwalia
Center of Innovative and Applied Bioprocessing
Mohali, Punjab, India

Amit Kumar Rai
Institute of Bioresources and Sustainable Development
Gangtok, Sikkim, India

Sunita Varjani
City University of Hong Kong
Kowloon, Hong Kong
and
University of Petroleum and Energy Studies
Dehradun, Uttarakhand, India

Raveendran Sindhu
Thangal Kunju Musaliar Institute of Technology
Kollam, Kerala, India

Parameswaran Binod
CSIR – National Institute for Interdisciplinary Science and Technology Trivandrum, Kerala, India

Saurabh Saran and Vinod Kumar
CSIR – Indian Institute of Integrative Medicine
Jammu, Jammu
and
Kashmir, India and Academy of Scientific and Innovative Research (AcSIR)
Ghaziabad, Uttar Pradesh, India

Ashok Pandey
Centre for Innovation and Translational Research
CSIR – Indian Institute of Toxicology Research
Lucknow, Uttar Pradesh, India

DOI: 10.1201/9781003355434-4

4.1 INTRODUCTION

Cellulose is considered one of the most abundant organic polymers on earth. It is a linear homopolysaccharide with a 1,4-glycosidic link that is composed of repeating β-D-glucopyranose units (Moon et al., 2011). Numerous organisms from the kingdoms of fungi, plants, and bacteria synthesize cellulose (Abeer et al., 2014; Thomas et al., 2018). Bacterial cellulose (BC) is composed of nanofibers that resemble ribbons woven together to form a three-dimensional network (Ruka et al., 2014). These nanofibers are highly crystalline, highly polymerized, and exceedingly pure (Choi and Shin, 2020). BCis free of the symbiotic components found in plant cellulose, such as hemicellulose, pectin, and lignin. After fermentation, the raw pellicle, which contains bacterial cells, leftover nutrients, and metabolic waste, can be easily separated from the BC network to make very pure products. According to (Tabuchi et al., 1998), in contrast with plant cellulose, BC generally possesses a higher degree of polymerization that varies depending on multiple factors, including bacterial genera/strains, nutrient sources, and various conditions of fermentation (Tahara et al., 1997).

These properties allow BC to be considered as a revolutionary wound-healing material make research on BC production technology very important. The main problem that restrains the use of BC as a wound-healing dressing is the ability to scale up production. Since the 1980s Johnson & Johnson has tried to produce BC commercially, predominantly as a lint to step up wound curing, although the data from several clinical trials has not been made public and the commercial products have not been formed (Ring et al., 1986, 1987). In the 1990s, a number of significant Japanese corporations and government agencies conducted an additional study for the commercial production of bacterial cellulose. Although $45 billion was spent for this research study, and various patents and publications resulted, no commercial production has consequently been realized. In the 1990s, fundamental research on BC biosynthesis in Poland was also pursued, and an effective *Gluconacetobacter* strain was discovered that can produce cellulose in more cost-effective nutrient media (Krystynowicz, 2005). Scientific research has focused on discovering new microbial producers of BC that would improve BC biosynthesis yield. (Gama et al., 2016) have scaled up production of BC as a hydrogel, with more than 500 tons capacity annually, and an estimated final cost of producing 1 kilogram of bacterial cellulose hydrogel of $25 and the payoff time of investment for this production is likely 4 years.

Research into improving BC yield has taken various directions: genetic construction and a search for innovative high-yield microbial producers (Gama et al., 2016; Sharma and Bhardwaj, 2019; Campano et al., 2015; Römling and Galperin, 2015); new technological approaches to cultivation (Sharma and Bhardwaj, 2019; Chen et al., 2018; Wang et al., 2019); and a search for alternative as well as cheap raw material sources (Molina-Ramírez et al., 2020a; Skiba et al., 2020; Velásquez-Riaño and Bojacá, 2017; Skiba et al., 2021). (Sharma and Bhardwaj, 2019) focused attention on high capital investment simultaneously with high functioning cost values as the key limitation to commercialization of BC at a sensible cost.

An issue emerges regarding the mode of cultivation, whether it is dynamic or stationary, because microbial producers of bacterial cellulose are obligate aerobes. Upgrading fermentation equipment is boosting the effectiveness of the BC production process. According to Choi et al. (2009), saccharified food waste was used as feedstock to produce a pilot-scale yield in a 50-L modified air-lift-type bubble column bioreactor under agitation. Stirred-tank bioreactors are thus utilized to increase production on a larger scale and produce more BC (Reiniati, 2017). The crystallinity index of such cellulose, however, is only 37%. Additionally, the use of cellulose produced under agitated conditions is restricted.

Agitation prevents the production of homogeneous, high-quality BC layers. This produces a fibrous or granular substance (Wu and Li, 2015).

This chapter discusses the most recent method of producing BC for industrial applications, and the rise of industrial production of BC. Nutritional parameters for BC synthesis and production-related issues are also described.

4.2 FERMENTATION STRATEGIES FOR THE PRODUCTION OF BACTERIAL CELLULOSE

Bacterial cellulose is mainly formed by two methods: agitated and static fermentation. The method selected depends on the application as the morphology and properties of BC produced by the two methods are quite different (Islam et al., 2017; Pang et al., 2020). In the case of static fermentation, a jelly-like pellicle forms at the air-liquid line of the culture medium. In contrast, in agitated fermentation, small, asymmetrical pellets are completely suspended within the culture fluid. In static fermentation, the cultured bacterial strain represents a privileged inherited constancy to constantly formed BC in maximum yield, but the efficiency of production is inadequate (Ross et al., 1991). Agitated fermentation is more efficient for industrial-scale production (Chen et al., 2018) but, according to Sani and Dahman (2010), it frequently causes a bacterial mutation that reduces output by producing a mutant that does not yield cellulose.

Although BC gels formed by the two methods show dissimilar macroscopic morphology, they retain the same three-dimensional reticulated microstructure (CPKelco Inc, 2020). Furthermore, Watanabe et al. (1998) found that BC produced in agitated culture has a lower degree of polymerization and is more crystalline than BC produced in static culture. The CP/MAS ^{13}C NMR study revealed that the cellulose Iα ratio in BC formed in agitated culture is less than in BC formed in static culture. Accordingly, the quantity of cellulose Iβ in BC attained in agitated culture is improved Watanabe et al. (1998). As a result, the two processes used to create BC do not produce the same mechanical properties. According to Krystynowicz et al. (2002), bacterial cellulose production has a larger Young's modulus than cellulose produced in an agitated culture. In contrast to BC produced in static culture, BC produced in agitated culture has a higher capacity to hold water and a higher viscosity of suspension.

4.3 MEDIUM FORMULATION FOR BACTERIAL CELLULOSE PRODUCTION

According to Andriani et al. (2020), a particular fermentation medium is composed of nitrogen and carbon sources and nutrient elements such as magnesium, potassium, phosphorus, and sulfur. The usual culture medium for the formation of BC was first described by Hestrin and Schramm (1954), and it contains 0.5% yeast extract, 0.5% peptone, 2.0% glucose, 0.115% citric acid, and 0.27% Na_2HPO_4 (in which glucose is used as a carbon source and peptone and yeast extract are used as nitrogen source). The medium pH is adjusted to 6.0 with NaOH or HCl. But the production cost is so high that it limits the industrial development of BC. Two approaches to overcome that cost are to boost the efficiency of production and to try to discover commercial nutrient sources as the cost of the culture medium accounts for approximately 30% of overall production cost (Rivas et al., 2004). There are now several approaches to support the production of BC:

1. Isolating novel bacterial strains from the natural environment that produce BC efficiently (Yang et al., 2013; Aydin and Aksoy, 2014).
2. Using genetic engineering techniques and conventional mutagenic techniques (such as chemical and ultraviolet mutagenesis) to screen high-production strains (Li et al., 2016).
3. Improve the conditions in the culture medium (Krystynowicz et al., 2002; Cheng et al., 2009).

Due to the high cost of the nutrients, conventional HS medium is expensive. According to (Keshk and Sameshima, 2005; Buldum et al., 2018), the bacteria feed with various sources of carbon (such as sucrose, glucose, arabitol, molasses, fructose, and mannitol) and different sources of nitrogen (such as peptone, corn steep liquor, and yeast extract). In the past few years, agricultural residues, as well as residues from other industries, including beet molasses, rotten fruit culture, coconut water or milk, waste beer yeast, liquid fermentation wastewater, and citrus fruit juice have been used as nutrient sources for the production of BC (Wu and Liu, 2013; Velásquez-Riaño and Bojacá, 2017; Cao et al., 2018; Julia et al., 2019). These nutrients drastically lower the production cost of BC and, further, alleviate environmental pollution induced by the inappropriate disposal of industrial waste.

The area required for fermentation in the production of BC is smaller than the area required to grow plants for the extraction of cellulose from wood. Also, the purification of BC is simplified because it is free of impurities compared with cellulose extracted from wood. Moreover, utilizing agricultural and industrial wastes in industrial fermentation not only lowers the cost, but also reduces pollution generated by waste (Hussain et al., 2019).

Static fermentation, which dates back to the 1970s in the Philippines, where it was used to produce nata de coco, was the first method utilized for the industrial production of bacterial cellulose, according to Iguchi et al. (2000). Nata de coco has a gelatinous morphology, a chilly, crispy flavor, and almost no cholesterol (Ullah et al., 2016). As a result, it became a popular raw food ingredient that is typically employed as a food and beverage preservative (Azeredo et al., 2019). The nata de coco thereafter became popular in Japan and south Asian nations, speeding up the industrial production of bacterial cellulose. Zhong started the industrial manufacturing of bacterial cellulose in China (1996). Hainan Yeguo Foods Co., Ltd. identified the *Gluconacetobacter xylinus* strain in 2020 from fermented coconut water and started the company to develop applications and produce bacterial cellulose. The China General Microbiological Culture Collection Center houses the *G. xylinus* strain 323 (No. 1186; Zhong, 2009). Particularly, these bacteria are amenable to the static fermentation that results in bacterial cellulose. Currently, this business has grown to become one of the biggest producers of bacterial cellulose goods in the world.

4.4 EFFECT OF NUTRIENTS ON THE SYNTHESIS OF BACTERIAL CELLULOSE: USING INEXPENSIVE SUBSTRATE TO PRODUCE BACTERIAL CELLULOSE

4.4.1 Different Types of Carbon Source for Bacterial Cellulose Production

To determine the fermentation medium's potential for producing bacterial cellulose using the *Komagataeibacter sucrofermentans* DSM 15973 bacterial strain, various sugars and crude glycerol were first evaluated. This bacterial strain failed to use xylose and lactose well, resulting in poor BC production of 1.7 g/L and 1.6 g/L, respectively. Ishihara et al. (2002) concluded that *Acetobacter xylinus* IFO 15606 turns xylose into D-xylonic acid. This makes xylonate build up in the broth and lowers the pH, which stops bacteria from growing and stops cellulose from being made. The lower concentration of BC seen in the lactose case can be attributed to Acetobacteraceae family members, of which *A. xylinus* is a part, not possessing the gene encoding the β-galactosidase (Battad-Bernardo et al., 2004). Compared to trials described in the literature (up to 4.1 g/L), glucose utilization led to decreased BC synthesis (1.2 g/L) (Son et al., 2003; Czaja et al., 2004). The use of increasing glucose concentration may result in the creation of gluconic acid as a byproduct during the biosynthesis of cellulose, which causes a steady drop in pH and lowers the yields of glucose from cellulose (Masaoka, et al., 1993). According to Dahman et al. (2010), the maximum concentration of BC (5.65 g/L) was reached in a media based on fructose, which is higher than the concentration of BC (2.06 g/L) attained in our work when fructose was used as the carbon source. The maximal concentration of BC is 3.2 g/L and 4.9 g/L for crude glycerol and commercial sucrose, respectively. The results are generally very consistent with works cited in the literature (Dahman et al. 2010; Mikkelsen et al. 2009). The highest quantities of BC are created when glucose (16%, w/w) and sucrose (32%, w/w) are employed as carbon sources, according to a recent study by Son et al. (2003).

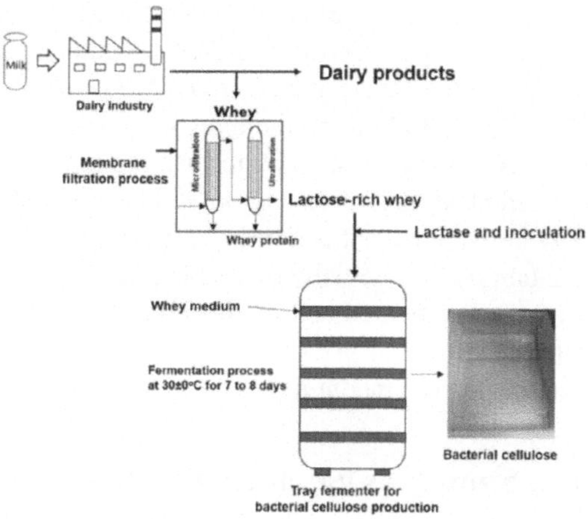

FIGURE 4.1 Utilization of whey for bacterial cellulose production.

4.4.2 Using Whey as a Substrate for the Synthesis of Bacteria Cellulose

In whey media devoid of any chemical additions, Kumar et al. (2021) evaluated the BC yield of a newly identified *Acetobacter pasteurianus* RSV-4 (MTCC 25117; Figure 4.1). An enzyme known as β-galactosidase (1.5 IU/ml) was utilized to break down whey lactose into its related monomers, glucose and galactose. As a result, *Acetobacter pasteurianus* specifically ate glucose for the formation of bacterial cellulose. After 8 days of microbial growth at 30°C, the maximum 5.6 g cellulose/L of whey concentration was reached, and the process was improved to manufacture bacterial cellulose on 120 L of whey medium under static conditions. Because it uses a waste byproduct (whey) from the milk-processing sector, the technique may be cost effective.

4.4.3 Bacterial Cellulose Synthesis from Fruit Residues

Komagataeibacter sucrofermentans DSM 15973 used citrus juices from whole oranges and grapefruits and watery extracts from citrus waste (peels) to produce BC. In comparison to lemon, grapefruit, and orange peel liquid extracts (5.2, 5.0, and 2.9 g/L, respectively), grapefruit and orange juice yield the most BC content (6.7 and 6.1 g/L, respectively). The BC produced from orange peel liquid extract has increased water holding capacity (26.5 g water/g, threefold greater), degree of polymerization (up to 6-fold higher), and crystalline record when compared to the cellulosic component recovered from depectinated orange peels. While evidence from the ^{13}C NMR spectra range revealed that I-α cellulose is the major form, absorption bands at 3240 and 3270 cm^{-1} in the infrared (IR) spectrum of BC show that the bacterial strain *K. sucrofermentans* produces both Iα- and Iβ-cellulose forms.

4.4.4 Other Agricultural Byproducts Utilized as Substrate for Bacterial Cellulose

Bacterial cellulose may be efficiently produced by using fermentation media made from the waste products of the biodiesel and confection industries. Using commercial sugar and unprocessed glycerol, batch fermentation with the bacterial strain *Komagataeibacter sucrofermentans* DSM (Deutsche Sammlung von Mikroorganismen) 15973 was initially carried out. The use of commercial sucrose (4.9 g/L) and unprocessed glycerol (3.2 g/L) led to the greatest amount of BC fixation. When crude glycerol and sunflower meal hydrolysate were combined as the only fermented media, 13.3 g/L of bacterial cellulose was produced. When flour-rich hydrolysates from waste streams from the confectionery sector were used, similar findings (13 g/L) were obtained. The characteristics of the BC produced using various fermented media included water holding capacities of 102–138 g(water)/g(dry), stress at break values of 72.3–139.5 MPa, viscosities of 4.7–9.3 dL/g, levels of polymerization of 1889.1–2672.8, and Young's modulus values of 0.97–1.64 GPa.

To lower production costs, BC has been made by fermenting agricultural and industrial wastes such as konjac powder, hemicellulose, textile, rice bark, and beer fermentation broth (Hong et al., 2012). Hong et al. specifically investigated cotton fabric waste as a substitute for producing BC. The wastes produced from cotton fabrics were initially treated with an ionic liquid and then saccharified using enzymes. When compared to untreated cotton, this cotton exhibits a 7-fold increase in enzymatic hydrolysis and fermented sugar; 10.8 g/L of BC were produced, which is around 83% more than what was obtained from the glucose medium.

4.4.5 Bacterial Cellulose Synthesis Is Influenced by the Nitrogen Source

Complex nitrogen sources caused BC production to rise when the concentration was higher. According to a study by Chai and Adnan (2018), BC synthesis decreased as the inorganic nitrogen supply fraction increased. The disruption of cell metabolism caused by an excessive nitrogen supply can limit the synthesis of microbial cellulose and lower nutrient uptake. Matsuoka et al. investigated the effect of lactate on BC synthesis in a different study (1996). BC synthesis was seen to be stimulated when 4% (w/v) lactate was added to maize steep liquor, yeast extract, or peptone as a nitrogen source and included fructose. Yim et al. (2017) discovered that BC textiles had a smooth surface (155.56 nm of roughness), 74.26 ± 5.24 % crystallinity, and 0.213 ± 0.01 mm of thickness when green tea and sucrose were used as the medium's nitrogen and carbon sources, respectively (Table 4.1).

TABLE 4.1
Production of Bacterial Cellulose Using Different Types of Substrates

Microbes	Medium Substrate	Yield g/L	Production Method	References
Gluconacetobacter sucrofermentans B-11267	Wheat thin stilage Whey	6.19 5.45	Fermentation under shaking conditions for 3 days at 30 °C	Revin et al. (2018)
Gluconacetobacter hansenii UCP1619	Corn steep liquor and glucose	9.63	Fermentation in static mode incubated at 30°C for 10 days	Costa et al. (2017)
Gluconacetobacter intermedius SNT-1	Cane molasses	12.60	Fermentation at 30°C, for 10 days at static conditions	(Tyagi and Suresh, 2015)
Acetobacter xylinum ATCC 23767	Tobacco waste extract	5.2	Two-stage fermentation (7 days and 16 days) at 30 °C with shaking 150 rpm	Ye et al. (2019)
Gluconacetobacter xylinus ATCC 10245	Potato peel	4.7	Fermentation at 35°C and pH 9.0 for 6 days in static conditions	Abdelraof et al. (2019)
Gluconacetobacter xylinus PTCC 1734	Beet molasses	4.56	Fermentation at 28°C, and pH 5.5 for 14 days in static conditions	Salari et al. (2019)
Gluconacetobacter medellinensis	Sugarcane and pineapple residues	3.24	Fermentation for 13 days at 28°C under static conditions	Algar et al. (2015)
Gluconacetobacter medellinensis ID13488	Apple residue/ sugarcane: 1/2.3	1.4	Fermentation for 14 days under static conditions	Urbina et al. (2017)

Microbes	Medium Substrate	Yield g/L	Production Method	References
Gluconacetobacter xylinus NRRL B-42	Grape bagasse	4.20	Fermentation for 14 days at 28°C under static conditions	Vazquez et al. (2013)
Acetobacter xylinum BPR2001	Molasses and corn steep liquor	3.01	Fermentation for 70 hours at 30°C under shaking conditions	Bae and Shoda (2005)
Komagataeibacter europaeus SGP37	Sweet lime pulp waste	6.30	Incubated in static conditions at 30°C for 16 days	Dubey et al. (2018)
Acetobacter xylinum NRRL B-42	Grape pomace extract/ corn steep liquor	6.7	Fermentation at 28°C, for 10 days in static conditions	Cerrutti et al. (2016)
Gluconacetobacter hansenii CGMCC 3917	Waste beer yeast treated with NaOH	2.40	Fermentation for 14 days in static conditions	Lin et al. (2014)

4.5 REACTOR DESIGN FOR BACTERIAL CELLULOSE PRODUCTION

A number of different kinds of reactors, such as static and agitated cultures, have been used in the production of BC. It is generally accepted that the amount of BC produced in a static culture is lower than that produced in an agitated culture. This is considered to be due to the slow growth of the microbes in the static culture and the limited oxygen supply. On the other hand, the appearance and rapid proliferation of cellulose-negative mutants in an agitated culture leads to a significant decrease in the amount of BC that is produced (Czaja et al., 2004). In a similar manner, the adhesion of the BC culture broth to the inner walls and uppermost regions of the vessel in the stir tanks of airlift reactors can reduce the amount of BC that is produced. This indicates the need for a reactor that can produce BC in a shorter amount of time and delay the formation of cellulose-negative sub-populations. In this context, a great number of attempts have been undertaken, and a wide variety of different kinds of reactors have been given some thought, with the goal of increasing the yield of bacterial cellulose.

4.5.1 Spinning Disc Reactor for the Production of Bacterial Cellulose

In this kind of reactor, almost half of the disc surface is submerged in broth and the other half is exposed to the atmosphere. Through constant rotation, the entire disc surface alternates between contact with the medium and air (Figure 4.2). BC-producing bacteria are attached to the surface of the disc, and get nutrients from the culture medium while submergedand atmospheric oxygen during the period the disc is exposed to air (Krystynowicz et al., 2000). This type of reactor's medium volume, rotational speed, and number of discs all have an impact on the amount of BC produced. According to the investigation by Krystynowicz et al., the maximum generation of BC was reached at 4 rpm and 0.71 cm^{-1} for the surface area to medium volume (S/V) ratio. A series of spherical discs are arranged on a horizontal shaft. These attached discs rotate continuously and are thus exposed to the fermentation broth and air. This type of reactor is advantageous since the bacteria that make BC don't have to contend with shear pressure and can just take oxygen from the surrounding air. In such a reactor *Gluconacetobacter* sp. has been used and generated up to 5.52 g/L of BCusing eight discs. Bacterial cellulose production was improved to 5.67 g/L when 1.25 vvm airing rate was maintained. The most favorable rotatory speed of the impeller was set at 15 rpm.

4.5.2 Spin-Filtration Reactor for the Production of Bacterial Cellulose

The use of a spin filter in the fermentation process was pioneered by Jung et al., (2007) who also investigated the production process. The fermentation process was carried out inside of a

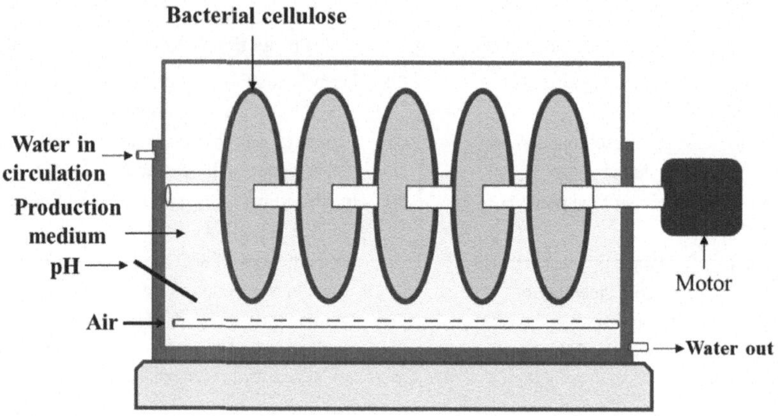

FIGURE 4.2 Layout of rotatory disc reactor for bacterial cellulose production.

fermenter that featured a turbine impeller with six flat blades and a spin filter that featured a cylinder that was covered with a stainless steel net. The lower portion of that piece of stainless steel was attached to the shaft of the agitator. At a temperature of 30°C and a medium with a pH of 5.0, *G. hansenii* PJK was used to produce BC. After 140 hours of growth in a periodic inoculation culture, the amount of BC that was produced reached its greatest concentration of 4.57 g/L. In comparison to the typical yield obtained from a traditional jar fermenter, that was an exceptionally high number.

4.5.3 Reactor with Silicone Membrane for Bacterial Cellulose Production

Bacterial cellulose pellicles develop in this type of static reactor on the exterior of an artificial membrane that is permeable to oxygen as well as on the surface of a liquid. In this kind of reactor, membrane roughness is the main factor influencing the amount of BC produced. This reactor increased the production of BC from *A. pasteurianus* AP-1SK, according to Yoshino et al. (1996). In a cylindrical vessel as opposed to a conservative static culture, the rate at which bacteria produced cellulose was up to twice as high. Therefore, a silicon air bag and a 100 mm thick silicon sheet were used to encase the lowest portion of the cylindrical vessel. Additionally, the use of a lustrous surface in that type of reactor resulted in a five times larger production of BC than that obtained from a marked surface.

4.6 CHALLENGES IN PILOT-SCALE PRODUCTION OF BACTERIAL CELLULOSE

Certain issues arise during the static fermentation of bacterial cellulose for industrial applications. Since the optimal temperature for *G. Xylinus* fermentation is 30°C (Son et al., 2003), the production of BC is restricted primarily to southeast Asian nations, including the Philippines, Thailand, Indonesia, and Vietnam. In the Chinese province of Hainan, heating is necessary to maintain a high temperature for the formation of BC in the winter, which undoubtedly increases energy consumption. Additionally, a high temperature frequently results in the contamination of food by bacteria, as well as other problems with production capacity, environmental pollution, and food safety. To resolve this challenge, Zhong and colleagues created a low-temperature domesticated method to lower the temperature at which bacteria ferment best to produce BC.

They also successfully screened out a low-temperature resistant strain to enable the production of BC at 10°C–20°C (Zhong, 2009). By using this method, it is now possible to produce BC in colder climates.

To reduce costs, industrial and agricultural wastes are employed as a source of nutrients for the synthesis of BC. Coconut water has continued to be one of the most significant nutritional sources for the industrial production of BC up to this time (Hainan Yeguo Foods Co., Ltd. 2020). Although coconut water was once an industrial waste, its enormous market demand has made supply insufficient, which has driven up the price. In place of coconut water, Zhong (2008a) constructed a two-step fermentation procedure. They carried out an earlier fermentation and used the results to produce BC from agricultural or industrial wastes such alcohol waste liquor, pineapple peels, corn cobs, citrus fruit juice, and apple juice. This procedure increases the nutrient resources available for the formation of BC, while producing a comparable yield to fermentation using coconut water. The agitated fermentation process has also been used successfully to produce industrial amounts of bacterial cellulose. Because of its unique three-dimensional reticulated network structure, it has been developed industrially as a suspending agent, specifically for particle suspension, according to Swazey (2014). After being freed from agitated fermentation, which is not suitable for transport or storage, the raw BC is filled with water. A multinational corporation, CPKelco Inc. (2020), developed natural hydrocolloids and fermented-based cellulose, also known as bacterial cellulose, through agitated fermentation. In the beginning, CPKelco commercially produced a soaked cake kind of BC that holds 10%–20% solid and the other balanced water (Swazey et al., 2013).

According to Beekmann et al. (2020), BC can be modified to meet specific medical needs, boost culture, and be loaded with useful active medicinal ingredients. This method is practical, effective, and transparent. Table 4.2 summarizes the several patents that are also available on the synthesis of BC.

TABLE 4.2
Patents on Bacterial Cellulose Production

S. No.	Application Year	Patent Number	Title	Application on Bacterial Cellulose	Inventors
1.	2018	CN109022534	Method for producing bacterial cellulose (BC) membrane using cornstalk and fuel cell from cornstalk degraded products	Producing a BC membrane using cornstalk	Song Jie, Lu Pengfei, Wu Zhen, Shen He, Xia Jun, Chen Duoli, Qian Xiaofeng, Shao Lixin, Jin Zhen, Zhang Zhiruo, Zhao Pusu, Xu Jiming
2.	2019	CN112092419	Preparation of BC composite board with tubular structure	The preparation method comprises BC composite material hydrogel into a tube	Yu Shuhong, Guan Qingfang, Han Zimeng, Yang Huaibin, Ling Zhangchi
3.	2018	CN108822323	Preparation method of BC membrane with moisturizing and sterilizing effects	In this invention purified BC membrane is submerged in 0.02%–0.4% tremella polysaccharide solution, the solution further removed and immersed in water and sterilized	Ma Xia, Wang Heng, Chen Shiwen

(Continued)

TABLE 4.2 (*Continued*)
Patents on Bacterial Cellulose Production

S. No.	Application Year	Patent Number	Title	Application on Bacterial Cellulose	Inventors
4.	2014	CN103773824	BC fermentation medium and method for producing BC using medium	BC fermentation medium with 25–35 g of cane sugar, 4–5 g of beef extract, 4–5 g of disodium hydrogen phosphate, 0.8–1.0 g of citric acid, 8–10 g of ethanol and the balance of hydrolysate of distillers' grains; pH value of the fermentation medium is 5.5–6.5	Zhang Wen, Zhang Minjuan, Li Yanjun, Liu Lan, Zhang Yan
5.	2019	CN109841899	Method for preparing three-dimensional network polymer gel electrolyte film based on BC	The method includes a step of heating a BC dry film in water to obtain a hydrogel film, a step of immersing the BC hydrogel film in an alcohol solvent	Wen Zhaoyin, Xu Dong, Wang Qing, Jin Jun, Wu Meifen
6.	2020	CN110894478	Culture medium for efficiently producing BC and production method of BC	The production of BC in the medium contains rice koji extract and distillers' grain culture medium base solution; and adding water and other nutrient substances	Mao Qing, Shui Chaoyi, Zhang Wei, Hu Xiulin, Wu Like
7.	2019	CN109825541	Method for regulating physical properties of cultured BC membrane by pH	The pH range in the industrial production of BC is favorably expanded; when the pH is in the range of 4–6, the lower the pH value is, the better the water absorption rate of the BC membrane	Lu Chaofeng, Lu Jianguo, Ding Haiwei, Wang Jiao, Cao Liming
8.	2016	CN105624231	Method for improving yield and quality of dry film by optimizing watermelon juice fermentation medium	The watermelon juice fermentation medium for fermenting BC; the yield is 6.39 g/L, 33.13% higher than the yield (4.80 g/L) of the BC dry film fermented in a basic medium	Chen Li, Lu Hongmei, Jiang Xiaolin, Li Jiayu
9.	2021	CN112876716	Method for rapidly preparing high-strength BC paper in large area	The method comprises the following steps of: (1) immersed in ethanol solution BC, (2) uniformly mixing the BC with a reinforcing agent, a thickening agent, a plasticizer and a wet strength agent, coating, drying, and separating the formed BC paper.	Wang Xiaohue, Ling Hao
10.	2021	CN112695009	Method for preparing bacterial cellulose in in situ fermentation	The method comprises the following specific steps: adding a 3.5 L (one gallon) gum raw material in a culture medium, further inoculate the culture medium with a selected strain for producing the BC, and incubation at 25°C	Huang Chao, Wu Yi, Hu Yong, Gao Jing, Li Zhixuan, Huang Zhongying

4.7 FUTURE PROSPECTS FOR THE PRODUCTION OF BACTERIAL CELLULOSE

Numerous industrial subsectors contribute daily to the generation of a substantial amount of garbage. There is a potential for a greater quantity of brewery, sugar, lignocellulosic, and other types of industrial waste to be converted into complex media for the manufacture of BC without the requirement of additional sources of nutrients, or into sources of carbon and nitrogen along with them. Agroindustrial wastes can be significantly utilized in the production of BC due to its extensive availability. This is particularly expected to be the case in developing nations. This is especially true for waste that does not require any complicated pretreatments, detoxification, or supplementing. The BC that is made from recycled media has a greater yield while maintaining the same physicochemical qualities as BC made from conventional media. It is crucial that BC be properly filtered because certain agro-waste might be colored and can absorb harmful substances. In light of the aforementioned factors, it is possible that the use of BC will be restricted in industries that are subject to stringent regulations, such as the biomedical, pharmaceutical, cosmetics, and food industries. The elimination of these industrial outputs will, from an ecological point of view, make it feasible to manage waste in an appropriate manner, hence lowering the risks that these wastes pose to both the environment and human health. This will be an effective and feasible solution to the problem of pollution.

4.8 CONCLUSIONS

Bacterial cellulose is an important biomaterial having wide application due to its desirable properties and structure. Many bacterial strains have been applied for the production of BC using different types of substrates. Selected strains have been studied for their ability to produce BC in various kinds of reactors, include under agitated and static conditions. BC finds potential applications in different types of industries, such as biomedical, foods, cosmetics, water purification, paper making, and many others. BC production is optimized using many substrates, including agroindustry byproducts. The utilization of agri-waste for bacterial cellulose production can help in minimizing both economic and environmental burdens imposed by different wastes generated in food industries. Further, recombinant technologies need to be explored more and can be applied for enhanced production of cellulose with better quality.

REFERENCES

Abdelraof, M., Hasanin, M. S., and El Saied, H. (2019). Ecofriendly green conversion of potato peel wastes to high productivity bacterial cellulose. *Carbohydr. Polym.* 211, 75–83.

Abeer, M. M., Mohd Amin, M. C., and Martin, C. (2014). A review of bacterial cellulose-based drug delivery systems: their biochemistry, current approaches and future prospects. *J. Pharm. Pharmacol.* 66, 1047–1061. doi: 10.1111/jphp.12234

Algar, I., Fernandes, S. C. M., Mondragon, G., Castro, C., Garcia-Astrain, C., Gabilondo, N., Retegi, A., and Eceiza, A. (2015). Pineapple agroindustrial residues for the production of high value bacterial cellulose with different morphologies. *J. Appl. Polym. Sci.* 132, 41237.

Andriani, D., Apriyana, A. Y., and Karina, M. (2020). The optimization of bacterial cellulose production and its applications: A review. *Cellulose* 27, 6747–6766. doi: 10.1007/s10570-020-03273-9

Aydin, Y. A., and Aksoy, N. D. (2014). Isolation and characterization of an efficient bacterial cellulose producer strain in agitated culture: *Gluconacetobacter hansenii* P2A. *Appl. Microbiol. Biotechnol.* 98, 1065–1075.

Azeredo, H. M. C., Barud, H., Farinas, C. S., Vasconcellos, V. M., and Claro, A. M. (2019). Bacterial cellulose as a raw material for food and food packaging applications. *Front. Sustain. Food Syst.* 3, 7.

Bae, S. O., and Shoda, M. (2005). Production of bacterial cellulose by Acetobacter xylinum BPR2001 using molasses medium in a jar fermentor. *Appl. Microbiol. Biotechnol.* 67, 45–51.

Battad-Bernardo, E., Mc Crindle, S. L., Couperwhite, I., and Neilan, B. A. (2004). Insertion of an *E. coli* lacZ gene in *Acetobacter xylinus* for the production of cellulose in whey. *FEMS Microbiol. Lett.* 231, 253–260.

Beekmann, U., Schmölzc, L., Lorkowskic, S., Werze, O., Thamma, J., Fischera, D., and Kralisch, D. (2020). Process control and scale-up of modified bacterial cellulose production for tailor-made anti-inflammatory drug delivery systems. *Carbohydrate Polymers* 236, 116062.

Buldum, G., Bismarck, A., and Mantalaris, A. (2018). Recombinant biosynthesis of bacterial cellulose in genetically modified *Escherichia coli*. *Bioprocess Biosyst. Eng.* 41, 265–279.

Campano, C., Balea, A., Blanco, A., and Negro, C. (2015). Enhancement of the fermentation process and properties of bacterial cellulose: A review. *Cellulose* 23, 57–91.

Cao, Y., Lu, S., and Yang, Y. (2018). Production of bacterial cellulose from byproduct of citrus juice processing (citrus pulp) by *Gluconacetobacter hansenii*. *Cellulose* 25, 6977–6988.

Cerrutti, P., Roldán, P., García, R. M., Galvagno, M. A., Vázquez, A., and Foresti, M. L. (2016). Production of bacterial nanocellulose from wine industry residues: Importance of fermentation time on pellicle characteristics. *J. Appl. Polym. Sci.* 133, 1–9.

Chai, J. M., and Adnan, A. (2018). Effect of different nitrogen source combinations on microbial cellulose production by Pseudomonas aeruginosa in batch fermentation. *IOP Conf. Ser. Mater. Sci. Eng.* 440, 012044.

Chen, G., Wu, G., Alriksson, B., Chen, L., Wang, W., Jönsson, L. J., and Hong, F. F. (2018). Scale-up of production of bacterial nanocellulose using submerged cultivation. *J. Chem. Technol. Biotechnol.* 93, 3418–3427. doi: 10.1002/jctb.5699

Cheng, K. C., Catchmark, J. M., and Demirci, A. (2009). Effect of different additives on bacterial cellulose production by *Acetobacter xylinum* and analysis of material property. *Cellulose* 16, 1033–1045.

Choi, C. N., Song, H. J., Kim, M. J., Chang, M. H., and Kim, S. J. (2009). Properties of bacterial cellulose produced in a pilot-scale spherical type bubble column bioreactor. *Korean J. Chem. Eng.* 26, 136–140.

Choi, S. M., and Shin, E. J. (2020). The nanofication and functionalization of bacterial cellulose and its applications. *Nanomaterials* 10, 406. doi: 10.3390/nano10030406

Costa, A. F. S., Almeida, F. C. G., Vinhas, G. M., and Sarubbo, L. A. (2017). Production of bacterial cellulose by *Gluconacetobacter hansenii* using Corn Steep Liquor as nutrient sources. *Front. Microbiol.* 8, 2027.

CPKelco Inc (2020). Available online at: https://www.cpkelco.com/products/fermentation-derived-cellulose-fdc/ (accessed August 20, 2020).

Czaja, W., Romanovicz, D., and Brown, R. M. (2004). Structural investigations of microbial cellulose produced in stationary and agitated culture. *Cellulose* 11, 403–411. doi: 10.1023/B:CELL.0000046412.11983.61

Dahman, Y., Jayasuriya, K. E., and Kalis, M. (2010). Potential of biocellulose nanofibers production from agricultural renewable resources: Preliminary study. *Appl. Biochem. Biotechnol.* 162, 1647–1659.

Dubey, S., Singh, J., and Singh, R. P. (2018). Biotransformation of sweet lime pulp waste into high-quality nanocellulose with an excellent productivity using Komagataeibacter europaeus SGP37 under static intermittent fed-batch cultivation. *Bioresour. Technol.* 247, 73–80.

Gama, M., Dourado, F., and Bielecki, S. (2016). *Bacterial Nanocellulose: From Biotechnology to Bio-Economy*. Cambridge, MA: Elsevier, p. 260.

Goelzer, F. D. E., Faria-Tischer, P. C. S., Vitorino, J. C., Sierakowski, M. R., and Tischer, C. A. (2009). Production and characterization of nanospheres of bacterial cellulose from Acetobacter xylinum from processed rice bark. *Materials Science and Engineering: C.* 29, 546–551. doi: 10.1016/j.msec.2008.10.013

Ha, J. H., Shehzad, O., and Khan, S. (2008). Production of bacterial cellulose by a static cultivation using the waste from beer culture broth. *Korean J. Chem. Eng.* 25, 812.

Hainan Yeguo Foods Co Ltd (2020). Available online at: http://www.yeguo.com (accessed August 20, 2020).

Hestrin, S., and Schramm, M. (1954). Synthesis of cellulose by *Acetobacter xylinum. II.* Preparation of freeze-dried cells capable of polymerizing glucose to cellulose. *Biochem. J.* 58, 345–352. doi: 10.1042/bj0580345

Hong, F., Guo, X., Zhang, S., Han, S. F., Yang, G., and Jönsson, L. J. (2012). Bacterial cellulose production from cotton-based waste textiles: Enzymatic saccharification enhanced by ionic liquid pretreatment. *Bioresour. Technol.* 104, 503–508. doi: 10.1016/j.biortech.2011.11.028.

Hussain, Z., Sajjad, W., Khan, T., and Wahid, F. (2019). Production of bacterial cellulose from industrial wastes: A review. *Cellulose* 26, 2895–2911.

Iguchi, M., Yamanaka, S., and Budhiono, A. (2000). Bacterial cellulose-a masterpiece of nature's arts. *J. Mater. Sci.* 35, 261–270. doi: 10.1023/A:1004775229149.

Ishihara, M., Matsunaga, M., Hayashi, N., and Tisler, V. (2002). Utilization of D-xylose as carbon source for production of bacterial cellulose. *Enzym. Microb. Technol.* 31, 986–991.

Islam, M. U., Ullah, M. W., Khan, S., Shah, N., and Park, J. K. (2017). Strategies for cost-effective and enhanced production of bacterial cellulose. *Int. J. Biol. Macromol.* 102, 1166–1173. doi: 10.1016/j.ijbiomac.2017.04.110

Julia, A., Andrea, C., Cláudio, G., Gloria, V., Emilia, S., and Leonie, S. (2019). Bacterial cellulose production using industrial fruit residues as subtract to industrial application. *Chem. Eng. Trans.* 74, 1165–1170.

Jung, J. Y., Khan, T., Park, J. K., and Chang, H. N. (2007). Production of bacterial cellulose by Gluconacetobacter hansenii using a novel bioreactor equipped with a spin filter, *Korean J. Chem. Eng.* 24, 265–271.

Keshk, S. M. A. S., and Sameshima, K. (2005). Evaluation of different carbon sources for bacterial cellulose production. *Afr. J. Biotechnol.* 4, 478–482. doi: 10.5897/AJB2005.000-3087.

Krystynowicz, A. (2005). *Method of Bacterial Cellulose Production, Sposob Wytwarzania Celulozy Bakteryjnej.* World Intellectual Property Organisation Poland. Patent PL WO2005003366A1, 13 January 2005.

Krystynowicz, A., Czaja, W., Pomorski, L., Kolodziejczyk, M., and Bielecki, S. (2000). The evaluation of usefulness of microbial cellulose as wound dressing material. 14th Forum for Applied Biotechnology, Gent, Belgium, Proceedings Part I, Meded Fac Landbouwwet-Rijksuniv Gent, 213–220.

Krystynowicz, A., Czaja, W., Wiktorowska-Jezierska, A., Goncalves-Miskiewicz, M., Turkiewicz, M., and Bielecki, S. (2002). Factors affecting the yield and properties of bacterial cellulose. *J. Ind. Microbiol. Biotechnol.* 29, 189–195. doi: 10.1038/sj.jim.7000303

Kumar, V., Sharma, D. K., Sandhu, P. P., Jadaun, J., Sangwan, R. S., Yadav, S. K. (2021, January). Sustainable process for the production of cellulose by an Acetobacter pasteurianus RSV-4 (MTCC 25117) on whey medium. *Cellulose* 28(1), 103–116.

Li, Y., Tian, J., Tian, H., Chen, X., Ping, W., Tian, C., et al. (2016). Mutation-based selection and analysis of *Komagataeibacter hansenii* HDM1-3 for improvement in bacterial cellulose production. *J. Appl. Microbiol.* 121, 1323–1334.

Lin, D., Lopez-Sanchez, P., Li, R., and Li, Z. (2014). Production of bacterial cellulose by Gluconacetobacter hansenii CGMCC 3917 using only waste beer yeast as nutrient source. *Bioresour. Technol.* 151, 113–119.

Masaoka, S., Ohe, T., and Sakota, N. (1993). Production of cellulose from glucose by *Acetobacter xylinum. J. Ferment. Bioeng.* 75, 18–22.

Matsuoka, M., Tsuchida, T., Matsushita, K., Adachi, O., and Yoshinaga, F. (1996). A synthetic medium for bacterial cellulose production by Acetobacter xylinum subsp.sucrofermentans. *Biosci Biotechnol Biochem.* 60, 575–579.

Mikkelsen, D., Flanagan, B. M., Dykes, G. A., and Gidley, M. J. (2009). Influence of different carbon sources on bacterial cellulose production by *Gluconacetobacter xylinus* strain ATCC 53524. *J. Appl. Microbiol.* 107, 576–583.

Molina-Ramírez, C., Álvarez, J., Zuluaga, R., Castro, C., and Gañán, P. (2020a). A novel approach using conventional methodologies to scale up BNC production using Komagataeibacter medellinensis and rotten banana waste as alternative. *Processes* 8, 1469.

Molina-Ramírez, C., Cañas-Gutiérrez, A., Castro, C., Zuluaga, R., and Gañán, P. (2020b). Effect of production process scale-up on the characteristics and properties of bacterial nanocellulose obtained from overripe Banana culture medium. *Carbohydr. Polym.* 240, 116341.

Moon, R. J., Martini, A., Nairn, J., Simonsen, J., and Youngblood, J. (2011). Cellulose nanomaterials review: Structure, properties and nanocomposites. *Chem. Soc. Rev.* 40, 941–3994. doi: 10.1039/c0cs00108b.

Pang, M. J., Huang, Y. H., Meng, F. S., Zhuang, Y., Liu, H., Du, M. L., et al. (2020). Application of bacterial cellulose in skin and bone tissue engineering. *Eur. Polym. J.* 122, 109365.

Reiniati, I. (2017). Bacterial Cellulose Nanocrystals: Production and Application. Electronic Thesis and Dissertation Repository. p. 4826.

Revin, V., Liyaskina, E., Nazarkina, M., Bogatyreva, A., Shchankin, M. (2018). Cost-effective production of bacterial cellulose using acidic food industry by-products. *Biotechnology and Industrial Microbiology* 49(1), 151–159.

Ring, D. F., Nashed, W., and Dow, T. (1986). Liquid Loaded Pad for Medical Applications. U.S. Patent No US4588400A, 13 May 1986.

Ring, D. F., Nashed, W., and Dow, T. (1987). Microbial Polysaccharide Articles and Methods of Production. U.S Patent No US4655758A, 7 April 1987.

Rivas, B., Moldes, A. B., Dominguez, J. M., and Parajo, J. C. (2004). Development of culture media containing spent yeast cells of *Debaryomyces hansenii* and corn steep liquor for lactic acid production with *Lactobacillus rhamnosus*. *Int. J. Food Microbiol.* 97, 93–98.

Römling, U., and Galperin, M. Y. (2015). Bacterial cellulose biosynthesis: Diversity of operons, subunits, products, and functions. *Trends Microbiol.* 23, 545–557.

Ross, P., Mayer, R., and Benziman, M. (1991). Cellulose biosynthesis and function in bacteria. *Microbiol. Rev.* 55, 35–58.

Ruka, D. R., Simon, G. P., and Dean, K. (2014). Harvesting fibrils from bacterial cellulose pellicles and subsequent formation of biodegradable poly-3-hydroxybutyrate nanocomposites. *Cellulose* 21, 4299–4308. doi: 10.1007/s10570-014-0415-z

Salari, M., Sowti Khiabani, M., Rezaei Mokarram, R., Ghanbarzadeh, B., and Samadi Kafil, H. (2019). Preparation and characterization of cellulose nanocrystals from bacterial cellulose produced in sugar beet molasses and cheese whey media. *Int. J. Biol. Macromol.* 122, 280–288.

Sani, A., and Dahman, Y. (2010). Improvements in the production of bacterial synthesized biocellulose nanofibres using different culture methods. *J. Chem. Technol. Biotechnol.* 85, 151–164. doi: 10.1002/jctb.2300

Sharma, C., and Bhardwaj, N. K. (2019). Bacterial nanocellulose: Present status, biomedical applications and future perspectives. *Mater. Sci. Eng. C.* 104, 109963.

Skiba, E. A., Budaeva, V. V., Ovchinnikova, E. V., Gladysheva, E. K., Kashcheyeva, E. I., Pavlov, I. N., and Sakovich, G. V. (2020). A technology for pilot production of bacterial cellulose from oat hulls. *Chem. Eng. Technol.* 383, 123128.

Skiba, E. A., Gladysheva, E. K., Golubev, D. S., Budaeva, V. V., Aleshina, L. A., Sakovich, G. V. (2021). Self-standardization of quality of bacterial cellulose produced by *Medusomyces gisevii* in nutrient media derived from Miscanthus biomass. *Carbohydr. Polym.* 252, 117178.

Son, H. J., Kim, H. G., Kim, K. K., Kim, H. S., Kim, Y. G., and Lee, S. J. (2003). Increased production of bacterial cellulose by *Acetobacter* sp. V6 in synthetic media under shaking culture conditions. *Bioresour. Technol.* 86, 215–219.

Swazey, J. M. (2014). *Surfactant Thickened Systems Comprising Microfibrous Cellulose and Methods of Making Same.* U.S. Patent No 8,772,359 B2. Washington, DC: U.S. Patent and Trademark Office.

Swazey, J. M., Morrison, N., Yang, Z. F., Compton, J., and Nolan, T. (2013). *Microfibrous Cellulose Composition Comprising Fermentation Media and Surfactant.* U.S. Patent No 10,292,927 B2. Washington, DC: U.S. Patent and Trademark Office.

Tabuchi, M., Watanabe, K., Morinaga, Y., and Yoshinaga, F. (1998). Acetylation of bacterial cellulose: Preparation of cellulose acetate having a high degree of polymerization. *Biosci. Biotechnol. Biochem.* 62, 1451–1454. doi: 10.1271/bbb.62.1451

Tahara, N., Tabuchi, M., Watanabe, K., Yano, H., Morinaga, Y., and Yoshinaga, F. (1997). Degree of polymerization of cellulose from *Acetobacter xylinum* BPR2001 decreased by cellulase produced by the strain. *Biosci. Biotechnol. Biochem.* 61, 1862–1865. doi: 10.1271/bbb.61.1862

Thomas, B., Raj, M. C., Athira, K. B., Rubiyah, M. H., Joy, J., Moores, A., et al. (2018). Nanocellulose, a versatile green platform: From biosources to materials and their applications. *Chem. Rev.* 118, 11575–11625. doi: 10.1021/acs.chemrev.7b00627

Tyagi, N., and Suresh, S. (2015). Production of cellulose from sugarcane molasses using Gluconacetobacter intermedius SNT-1: Optimization & characterization. *J. Clean. Prod.* 112, 71–80.

Ullah, H., Santos, H. A., and Khan, T. (2016). Applications of bacterial cellulose in food, cosmetics and drug delivery. *Cellulose* 23, 2291–2314.

Urbina, L., Hernández-Arriaga, A. M., Eceiza, A., Gabilondo, N., Corcuera, M. A., Prieto, M. A., and Retegi, A. (2017). By-products of the cider production: An alternative source of nutrients to produce bacterial cellulose. *Cellulose* 24, 2071–2082.

Vazquez, A., Foresti, M. L., Cerrutti, P., and Galvagno, M. (2013). Bacterial Cellulose from simple and low cost production media by Gluconacetobacter xylinus. *J. Polym. Environ.* 21, 545–554.

Velásquez-Riaño, M., and Bojacá, V. (2017). Production of bacterial cellulose from alternative low-cost substrates. *Cellulose* 24, 2677–2698.

Wang, J., Tavakoli, J., and Tang, Y. (2019). Bacterial cellulose production, properties and applications with different culture methods—A review. *Carbohydr. Polym.* 2019, 63–76.

Watanabe, K., Tabuchi, M., Morinaga, Y., and Yoshinaga, F. (1998). Structural features and properties of bacterial cellulose produced in agitated culture. *Cellulose* 5, 187–200.

Wu, J. M., and Liu, R. H. (2013). Cost-effective production of bacterial cellulose in static cultures using distillery wastewater. *J. Biosci. Bioeng.* 115, 284–290.

Wu, S.-C., and Li, M.-H. (2015). Production of bacterial cellulose membranes in a modified airlift bioreactor by *Gluconacetobacter xylinus. J. Biosci. Bioeng.* 120, 444–449.

Yang, Y., Jia, J., Xing, J., Chen, J., and Lu, S. (2013). Isolation and characteristics analysis of a novel high bacterial cellulose producing strain *Gluconacetobacter intermedius* CIs26. *Carbohydr. Polym.* 92, 2012–2017.

Ye, J., Zheng, S., Zhang, Z., Yang, F., Ma, K., Feng, Y., and Yang, X. (2019). Bacterial cellulose production by Acetobacter xylinum ATCC 23767 using tobacco waste extract as culture medium. *Bioresource Technology* 274, 518–524.

Yim, Su Min, Eun Song Ji, and Rim Kim Hye (2017). Production and characterization of bacterial cellulose fabrics by nitrogen sources of tea and carbon sources of sugar. *Process Biochemistry* 59, Part A, 26–36.

Yoshino, T., Asakura, T., and Toda, K. (1996). Cellulose production by *Acetobacter pasteurianus* on silicon Membrane. *J Ferment Bioeng.* 81, 32–36. doi: 10.1016/0922-338X(96)83116-3

Zhong, C. (1996). *Method for Edible Fiber Product by Fermentation of Coconut Juice.* Chinese patent No: CN1066926C. Haikou: hina National Intellectual Property Administration.

Zhong, C. (2008a). *Compressed Coconut and Method of Preparing the Same.* Chinese patent No: CN101278737A. Haikou: China National Intellectual Property Administration.

Zhong, C. (2009). *Gluconacetobacter Oboediens Strain and Method for Breeding and Producing Bacteria cellulose Thereof.* Chinese patent No: CN101608167. Haikou: China National Intellectual Property Administration.

5 Downstream Process for Bacterial Cellulose Production

Júlio Cesar de Carvalho, Luiz Alberto Junior Letti, Leonardo Wedderhoff Herrmann, Alexander da Silva Vale, Walter José Martínez-Burgos, Rafaela de Oliveira Penha, Thamarys Scapini, and Carlos Ricardo Soccol
Federal University of Paraná
Curitiba, Paraná, Brazil

5.1 INTRODUCTION

Bacterial cellulose (BC) is an exopolysaccharide composed exclusively of β-1,4-glucopyranosyl units that form a biopolymer. Different from plant cellulose, this polymer is originated from bacterial cells, mainly from the genus *Glucoacetobacter*, formerly *Komagataeibacter* or *Acetobacter*, in species such as *Gluconacetobacter xylinus* and *Gluconacetobacter hansenii*. Most of the current BC production is performed at laboratory scale utilizing glucose, peptone, yeast extract, and synthetic media; however, there is intense research into alternative carbon and nitrogen sources since BC is classified as a green product. The BC market was estimated in 2020 to be US $506.8 million, and it is expected to reach US $980.9 million by 2028, with a growth rate of 8.1% each year. Such values represent the crescent interest for the technology in the recent decades (Cacicedo et al. 2016; Chawla et al. 2009; Biospace 2022).

Similar to most bioproducts, BC needs to undergo purification processes, also called downstream processes, before being used for any application. After being produced through fermentation, whether using conventional or renewable substrates, BC must be separated from the bacteria and the culture medium. The downstream strategy determines several properties of the final product and its application. As an extracellular biopolymer, most of the cellulose gathers in a biofilm, which is generally simple to be purified through unit operations such as filtration or centrifugation. The resulting BC still presents remaining bacteria due to the complex polymer structure, therefore requiring a decontamination step. Finally, the cellulose retains a large amount of water in its structure, which is usually removed using a drying process. All of those processes characterize a generic purification of the BC (Ebrahimi et al., 2016; Forte et al., 2021).

It is important to know the physicochemical characteristics of BC before choosing any downstream operation. Although chemically equal to plant cellulose, the BC presents significant structural differences, primarily due to the smaller size of its fibrils. The bacterial polymer has around 2 to 4 nm diameter fibril building blocks, which are grouped as ribbons of 20 to 120 nm width and 1 to 9 μm length, which is around 1% of the size of wood fibers (Ebrahimi et al., 2016; Skiba et al., 2020; Cacicedo et al., 2016). The tensile strength of the biopolymer reportedly ranges from 40 to 450 MPa, the Young's modulus from 5 to 30 GPa (Eichhorn et al., 2010), and the water absorption capacity is high, being capable of absorbing between 0.1 and 400 times its dry weight of water (Moon et al., 2011). Besides, the density of BC after drying varies from 0.05 to 0.16 g cm^{-3} (Zeng et al., 2014). Other interesting physical properties regarding mechanical stability are high thermostability and ultrafine fibrous structure (Wang et al., 2019), as well as high

 DOI: 10.1201/9781003355434-5

biocompatibility, which indicates no biological toxicity and low or no immune response against it (Cacicedo et al., 2016).

One of the main advantages of BC compared to plant polymers is that it does not present hemicellulose or lignin in its composition, both of them being major recalcitrant products in plant biomasses that are difficult to degrade and hinder several cellulose products along time (Gea et al., 2011). However, when produced by bacteria, some major drawbacks are that the cellulose fibrils are relatively poorly connected to the matrix and inappropriately oriented. Those characteristics are the reason for testing chemical treatments with coupling agents such as silane and testing strategies for cultivation in silicone tubes, respectively (Ebrahimi et al., 2016; Putra et al., 2008).

Besides, changes in the fermentation process can drastically modify the BC's porosity, thickness, strength, and other characteristics. For instance, different carbon sources such as sucrose, glucose, fructose, and glycerol, as well as the inoculum concentration and cultivation time, significantly alter the BC configuration. Usually, the film porosity decreases with increasing film thickness (Al-Shamary and Darwash, 2013). Physical properties are also highly dependent on the type of fermentation strategy used, whether the process is performed in batch, batch-fed, or continuous modes, or even under agitated or static conditions (Cacicedo et al., 2016). For example, when the process is static, cellulose is obtained in the form of a three-dimensional interconnected reticular film and when the process is carried out under agitated conditions, cellulose particles similar to spheres of irregular shape, pellets, or irregular masses are produced (Esa et al., 2014).

Besides the fermentation strategies, which final product the BC will be applied in is also determinant for the downstream selection. When destined to encapsulate material for enzymatic reactions, membranes used for separation, packaging, or even electronics purposes, the purity is not a strict requirement, but rather the material's mechanical properties. Yet, when applied in human food, such as in the dessert called *nata de coco*, or for human health products, such as covering wounds and burn treatments or, more recently, in medical tissue engineering, the BC product must be free from contamination from the bacteria or fermentation broth. Much of the BC applications also require further steps to reach a final product, such as adding salts or chemical agents, or combining with other polymers (Cacicedo et al., 2016).

Another obstacle found during the purification of BC, especially when destined for medical applications, is that virtually all bacteria with high BC productivity are classified as Gram-negative, a bacterial group that produces membrane lipopolysaccharides (LPS) that act as endotoxins. These LPS molecules are thermotolerant; removing them can take days, requiring complex and aggressive processes (Ávila et al., 2014). Furthermore, thicker BC films are preferred for matrices implants, but conventional purification methods for the polymer present lower efficiency even when compared to thin films (Pigaleva et al., 2019; Gromovykh et al., 2018). Regarding the maximum LPS concentration allowed, products destined for human consumption should have a concentration of endotoxins lower than 0.5 endotoxin units (EU) per milliliter (Ávila et al., 2014; Pigaleva et al., 2019).

Therefore, both the characteristics of the BC after fermentation and the desired features in the final product guide the development of an optimal downstream process. This chapter provides an overview of the general purification steps taken for obtaining BC reported in the literature. Some of the downstream processes are presented and compared in Table 5.1.

5.2 ISOLATION OF BACTERIAL CELLULOSE FROM THE CULTURE MEDIUM

The first downstream unit operation after BC fermentation is to separate the polymer from most of the fermentation medium and the bacteria biomass. According to the BC characteristics, the main operations involved in the separation are filtration, centrifugation, and, in some cases, the polymer is separated just by gathering and picking the biofilm over the solution (Table 5.1).

TABLE 5.1
Examples of Bacterial Cellulose (BC) Downstream Processes

Strain	Fermentation Process	Separation	Decontamination	Dryer	Application	References
Gluconacetobacter xylinus ATCC 53524	Static	Cellulose pellicle was harvested by hand	Incubate the BC in 0.1 M NaOH solution at 90°C for 30	Cellulose pellicle was lyophilized at 10°C and 0.006 mbar	Not evaluated (NE)	(Mikkelsen et al., 2009)
Gluconacetobacter xylinus	Stirred-tank bioreactor	Centrifugation	Incubate the BC in 0.1 M NaOH solution at 90°C for 30–120 min	Cellulose pellicle was frozen at −80°C and lyophilized	NE	(Reiniati, Hrymak, and Margaritis, 2017)
Gluconacetobacter hansenii UCP1619	Static	Cellulose pellicle was harvested by hand	Incubate the BC in 0.1 M NaOH solution at 90°C for 20 min; after the process, the BC membranes were irradiated with gamma rays	Partial drying using pressure	Prototype of a face mask	(Amorim et al., 2022)
Gluconacetobacter xylinus	Static	Cellulose pellicle was harvested by hand	Incubate the BC in 2% NaOH solution for 1 h; subsequently, the BC has immersed in 1% NaClO solution for 30 min; BC was washed with water and sterilized at 120°C for 15 min	NE	Developing a composite based on BC and collagen for tissue engineering applications	(Saska et al., 2012)
Gluconacetobacter hansenii GH-1/2008	Static	Filtration	The BC was immersed in a solution containing 5% sodium dodecyl sulfate (SDS) The BC was immersed in a solution of sodium bicarbonate The BC was immersed in a RIPA buffer (25 mM Tris-HCL, pH 7–8, 150 mM NaCl, 0.5% sodium deoxycholate, 1% Triton X-100, 0.1% SDS) Processing in media containing CO_2 under high pressure	Purified films were frozen at −80 °C in liquid nitrogen and subjected to lyophilization	Biomedical applications	(Pigaleva et al., 2019)
Gluconacetobacter xylinus	Static	Cellulose pellicle was harvested by hand	Incubate the BC in 0.3 M NaOH solution and sterilize it in an autoclave	Cellulose pellicle was dried in an oven at 50°C for 24 h or lyophilized	Drug delivery system	(Badshah et al., 2018)

Gluconacetobacter hansenii ATCC 23769	Static	Cellulose pellicle was harvested by hand	Incubate BC in 1% NaOH solution at 95°C for 60 min	The BC pellicle was oven-dried at 50°C on a silicon surface for 48 h Frozen at −30°C and then lyophilized	NE	(Vasconcellos and Farinas, 2018)
Gluconacetobacter xylinus (GX) ATCC 11142 *Gluconacetobacter europeaus* (GE) MF03	Static	Cellulose pellicle was harvested by hand	The BC was incubated in ethanol solution; subsequently transferred to deionized water and boiled for 40 min, four times more with 0.1 M NaOH at 90 °C for 20 min	The BC pellicle was kept at room temperature for 4 days The BC was frozen in liquid nitrogen for 5 min; then the sample was lyophilized at −80°C and 0.005 mbar for 12 h Supercritical CO_2 drying	NE	(Zeng, Laromaine, and Roig, 2014)
Gluconacetobacter xylinus 0416	Static	Cellulose pellicle was harvested by hand	The BC pellicle formed was harvested, washed with distilled water, and purified by immersing into 0.1 M NaOH aqueous solution at 80 °C for 1 h	The BC pellicle was dried at 40 °C	NE	(Azmi et al., 2021)
Gluconacetobacter xylinus	Static	Cellulose pellicle was harvested by hand	Incubate the BC in a solution of 0.5 M NaOH at 90°C for 1 h	Air-drying oven at 80°C or by freeze drying	NE	(Al-Shamary and Darwash, 2013)
Gluconacetobacter xylinus	Static	Cellulose pellicle was harvested by hand	BC was immersed overnight in 2.5 wt.% NaOH	NE	Paper production	(Gea et al., 2011)

5.2.1 Filtration

Filtration is a unit operation mainly applied for separating solids and semisolids from a liquid solution. This is one of the most used operations for separating BC from the fermentation medium, especially when the polymer production process was carried out under agitated conditions. The type of agitation and its intensity during growth is responsible for the round-like form that the BC adopts, and the diameter of the spheres is a function of the agitation speed in the fermentation. For instance, when the process is carried out at speeds of 125 rpm, cellulose spheres reach an average diameter of 8 mm. However, when fermentation is carried out at higher stirring speeds, such as 150 rpm, the diameter decreases to 2 mm (Hu et al., 2013). To separate the BC from the medium, the process operating conditions must be previously known and, thus, properly selecting the diameter of the filter is crucial. The filter media used for this separation are different porosity Whatman paper and cotton gauze (Moniri et al., 2017; Castro et al., 2011; Matsuoka et al., 1996). The processes can be carried out under vacuum or by gravity depending on the filtering medium. Gravity systems are the most recommended, as they affect the characteristics of BC to a lesser extent.

5.2.2 Centrifugation

Centrifugation is a unit operation applied to separate complex heterogeneous mixtures of solids and liquids or between liquids. As BC is denser than the fermentation medium, centrifugation is a suitable operation for separating the biopolymer. However, the centrifugation conditions are important to be evaluated since the speed can affect the water holding capacity of the biopolymer. For example, a centrifuge speed of approximately 1500 *g* can reduce the amount of water kept in the polymer by up to 75% when BC is produced under static conditions. Similarly, when the polymer is produced under nonstatic conditions, the separation using a centrifuge speed of 1500 *g* can reduce up to 98% the amount of water in the polymer. Further investigation showed that rehydration did not restore 100% of the water holding capacity. Although standard centrifugation conditions are not stipulated, this must be carried out at the lowest possible speed as these processes negatively affect the BC water retention capacity, regardless of the method used for their production (Krystynowicz et al., 2002; Masoodi et al., 2012; Liebner et al., 2010). Figure 5.1 shows the effect of centrifugation speed on the water holding capacity of BC produced under different conditions.

5.2.3 Manual Harvesting

As the BC is capable of forming a biofilm over the fermentation broth when cultivated statically, one of the possible separation techniques is to gather the polymer film "manually" and harvest it

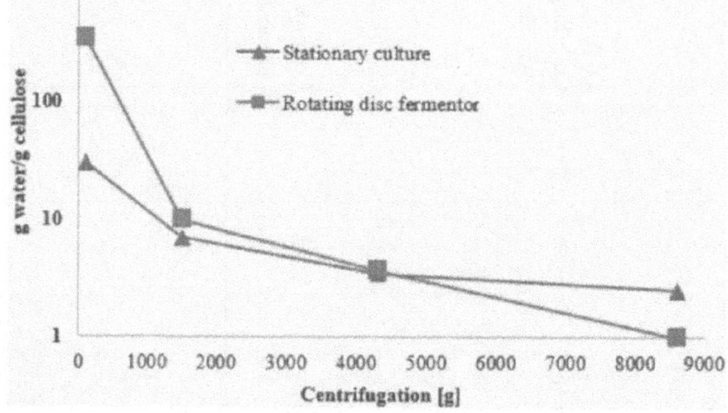

FIGURE 5.1 Water-holding capacity after centrifugation of bacterial celluloses originated by different types of bioreactor and destined to centrifugation right after fermentation.

Source: Krystynowicz et al., (2002).

with clamps, spatula, rotating impellers or similar instruments, as well as of sieves. Actually, this separation mechanism is the most popular for BC production, especially at laboratory and small scales (Table 5.1). When using automatic devices, it is recommended to use low agitation, up to 4 rpm, to improve the polymer's yield and water retention capacity (Krystynowicz et al., 2002). The main advantages of "manual" separation compared to centrifugation and filtration is that it does not affect the mechanical properties and water holding capacity of BC, and provides an alternative with reduced energy cost for the industrial production of the polymer. However, as a manual strategy, little to no information is available in terms of operational conditions.

5.3 DECONTAMINATION

After the separation of BC from the fermented broth, it still requires purification, since microorganisms, proteins, lipids, toxins, and other constituents originated from the culture medium may be adhered to the cellulose microfibrils (Azeredo et al., 2019). The decontamination or sterilization methods are critical steps in the BC purification process, especially when the films are used in the development of new products destined for human consumption, such as cosmetics, biomaterials, food, and drugs, requiring high quality control (Cacicedo et al., 2016). In general, conventional decontamination processes are performed in three main steps, the first being an initial washing of the BC with water to remove nutrients and proteins from the fermentation broth, the second being chemical treatment, and the third neutralization of the chemicals used, and successive washing with distilled water (Moniri et al., 2017). The most frequent strategy applied in the chemical treatment is the use of alkaline solutions, mainly sodium hydroxide (NaOH), to remove the cells (Table 5.1), yet some new strategies using, for instance, supercritical carbon dioxide (CO_2) are being developed.

5.3.1 Alkaline Treatment

The alkaline treatment with NaOH has been widely employed in the process of decontamination of BC (Table 5.1). The high pH values favor the lysis of microbial cells and the removal of cell debris. For plant biomass, the use of NaOH is generally associated with an efficient separation of the cellulosic fraction from hemicellulose and lignin during the pretreatment. These types of biomass present more complexity than BC, which does not present either hemicellulose or lignin, or other polysaccharides associated with its structure, making the alkaline treatment an optimal strategy for the purification process (Moniri et al., 2017; Procentese et al., 2017). This treatment is relatively simple and consists of immersing the BC in a NaOH solution. As shown in Table 5.1, the concentration of the alkaline solution, temperature, and time used in this process, for example 0.1–0.3 M, 90°C–95 °C and 20–120 min, respectively, do not differ from much of the literature examples.

Other compounds such as sodium dodecyl sulfate (SDS), sodium hypochlorite (NaClO), ethanol, and acetic acid can be used for the same decontamination purpose, each one having different mechanisms to eliminate cells (Chawla et al., 2009; Zeng et al., 2014). Successive washing with reverse osmosis water or hot water is also an alternative to performing BC decontamination (Azeredo et al., 2019). These methods can be used separately or combined to reach better sterilization. However, it is important to standardize both BC film purification and sterilization processes.

The efficiency of the washing process and the number of cycles required to remove soluble compounds can be determined using spectrophotometric techniques. For instance, the presence of insoluble nutrients can be monitored by reading the water absorbance at different wavelengths after washing, while proteins can be quantified using the Bradford method. A study conducted by Reiniati et al. (2017) using both strategies showed the necessity of three water washing steps using 2.5 times the fermentation volume to reach significant removal of nutrients from the culture medium, and four cycles for protein removal. Once these compounds were removed, the samples were chemically treated, neutralized, and dried, producing microorganism-free cellulose biofilms.

After these steps, scanning electron microscopy can be used to verify both the physical structure of BC and the presence of microbial cells, being a resourceful tool (Amorim et al., 2022; Reiniati et al., 2017). However, this method is not efficient to determine the presence of metabolically active cells, since the three-dimensional network of cellulose can entrap BC-producing microorganisms and hinder their visualization, as well as hindering the contact between cells and chemical agents and their removal through traditional purification techniques. For instance, Amorim et al. (2022) showed that the treatment with 0.1 M NaOH solution at 90°C for 20 min was not enough to sterilize BC, since after incubation of this cellulose film in a culture medium at 30°C for 7 days the bacterial growth reached 3.7×10^4 CFU. Therefore, after alkaline decontamination, the authors chose to perform an additional step of BC sterilization by gamma ray irradiation. None of the evaluated samples exhibited bacterial growth after combining both processes, indicating a promising methodology for producing sterile cellulose films (Amorim et al., 2022).

5.3.2 Supercritical Carbon Dioxide Treatment

To develop BC-based products and meet the market demand, a rapid methodology for simultaneous purification and sterilization of thick films with little to no degradation of the polymer was proposed by Pigaleva et al. (2019). The authors used processing in a CO_2-containing medium under high pressure and compared the results with the conventional method. The results showed that the application of supercritical fluid could render the BC samples sterile, that is, without the presence of viable microbial cells. However, atomic force microscopy (AFM) analyses showed that cell membranes were destroyed, though not completely. The concentration of endotoxins was reduced, yet it was still higher than permitted for medical applications. One possible explanation is that the endotoxins were partially degraded and spread throughout BC, presumably by forming hydrogen bonds with the cellulose fibrils. According to the authors, new methods, involving a two-phase system under high pressure (CO_2 and other polar media) should be devised to achieve the desired level of purification, although the use of supercritical CO_2 adds a significant cost to the downstream process (Pigaleva et al., 2019).

5.4 DRYING

As previously mentioned, one of the properties of BC is its high water absorption capacity. The amount of moisture it can absorb varies from 0.1 to 400 times the polymer dry weight. However, the more water is present in the compound, the weaker the mechanical properties of the material. For instance, reducing the humidity from 60% to 30% enhanced ten times the tensile modulus and two times the tensile strength. In another experiment, the thickness increased 2.4 times when dry cellulose films entered in contact with water indicating that the properties could be reversible according to the water content (Ebrahimi et al., 2016; Moon et al., 2011; Masoodi et al., 2012). As most sterilization strategies are followed by a washing step to remove the excess of alkaline agent, the water near the end of the downstream process is high, requiring a drying and concentration step. The majority of the BC drying operations are simple, either performed at room temperature or drying ovens, natural or reduced pressures, yet they can also be freeze dried or spray dried.

5.4.1 Room Temperature Drying and Oven Drying

One of the most inexpensive and less labor-requiring techniques is to simply lay the BC to dry at room temperature, most of the times between 20°C and 30°C. This process relies on the natural evaporation of the water contained in the polymer. Drying at room temperature demands more time than the heated options, such as 2 or 3 days; however, one of its great advantages is maintaining the general structure of the cellulose intact (Ebrahimi et al., 2016).

Water removal through increased temperatures, in contrast, accelerates the process. It can be performed in drying ovens with or without circulating air, for instance in temperatures of 50°C to 100°C, while the higher the temperature, the faster the process. However, high temperatures such as 100°C can degrade the cellulose structure, reducing tensile strength (Ebrahimi et al., 2016) and lowering the porosity of the polymer. The BC shrinks due to the partial collapse of some of its fibrils, reducing the polymer size and thickness. On the other hand, the thermal stability and crystallinity remain the same (Al-Shamary and Darwash, 2013; Vasconcellos and Farinas, 2018). These characteristics can be interesting depending on the application of the compound, with relatively low cost in terms of energy and labor. Vacuum drying can also greatly accelerate the process.

5.4.2 Dewatering by Pressing Plates

Submitting BC biofilms to pressing between metallic plates or devices and increasing pressure can also remove water content from the polymer, regardless of the plate's format and configuration. The pressure systems allow an accelerated removal of the polymer moisture in the first steps of drying and can be followed by oven or room temperature drying to diversify the product's final characteristic and optimize both energy and time consumption. Although the BC porosity decreases when using pressing devices, both tensile strength and tensile modulus increase, turning the fibers wider and favoring the stabilization of internal bonds (Ebrahimi et al., 2016; Gea et al., 2011).

5.4.3 Lyophilization

The freeze-drying technique, also called lyophilization, is a process that removes water from a sample by freezing it and then sublimating it under reduced pressure. This operation can conserve more of the polymer's original porosity and thickness compared to oven drying. The very structure of the product is different to that obtained by drying pressed BC films: the lyophilized material may have an apparent density as low as 8 kg/m^3, with high porosity and hydration capacity, besides keeping the crystallinity and thermal stability. Although lyophilizing maintains the integrity of internal bonds and fibrils, the process reduces the transparency and confers a whitish color to BC, which is not desirable in some products. In addition, depending on the case, freeze drying can represent a higher energy cost and difficulties to scale up to industrial scale (Al-Shamary and Darwash, 2013; Vasconcellos and Farinas, 2018).

5.4.4 Spray Drying

One of the most versatile processes applied for industrial drying is the use of spray dryers. This technique and equipment consist in forcing the liquid-solid mixture through small gaps called atomizers, and subsequently submitting the droplets to heated air for quick drying and producing powder-like products. The spray dryer can also be applied for drying BC, reaching faster removal of water content and relatively controlled characteristics. However, it cannot be applied directly to whole biofilms, but it may be adequate for drying the product clumps obtained in agitated reactors or, for instance, a homogenized solution of 1% shredded and chopped BC film. When submitted to this strategy, the resulting powder presented more spherical granular particles, higher density, lower porosity, and higher flowability than the lyophilized BC. The high temperatures applied in the inlet and outlet gas flow also increase the sterilization of the biopolymer, eliminating possible remaining cells from its structure or even exempting the sterilization step (Amin et al., 2014; Peng et al., 2012).

5.5 PATENTS AND INNOVATIONS

Although the first reports of BC production date back to the 1980s, there has been a growing interest in BC technology in recent years as a promising technology for polymer production and application, which also extends to its downstream process. Most of the patented documents focused on this technology intended to protect producing steps of the process or the several types of products wherein BC can be applied. However, few patents present detailed information of the downstream processes used for the polymer, especially in the decontamination step, but can indicate which obstacles are being researched for industry application.

For example, patent number CN107582509-A (Liu and Shi, 2018) protects a new strategy for BC decontamination and efficient removal of endotoxins, which maximum concentration accepted is 0.5 EU/mL. The inventors added a solution of Chinese medicine extract, citric acid, bactericide, and vitamins to the biofilm and used ultrasonic mixing for 30 min. The method could reduce the toxin content to 0.2 EU/mL, allowing the BC film to be applied for preparing a facial mask, also protected in the document.

Some of the patents also combine strategies for decontamination, mixing different chemical agents and modifying the number of washings. Patent number CN113980185-A (Zhang et al., 2022) uses 3 to 4 sequential deionized water washings followed by hydrolyzing with sulfuric acid, dialyzing and adding 1 M Tris buffer for stabilizing. This technique results in an amorphous BC that can be used for a hydrogel dressing with silver particles with antibacterial activity. Another document, CN113881085-A (J. Liu et al., 2022), used a hot alkaline decontamination, followed by washings with lye and hydrogen peroxide. The resulting BC membrane was tested with mammalian cells and reached excellent biocompatibility. These innovations indicate the use of multiple chemicals combined and several washing steps to purify the BC with the maximum quality primarily for medical purposes.

Interestingly, some strategies are focused on separating the cellulose fibrils as much as possible to formulate hydrogels, not interested in the natural biofilm formation. For instance, patent number CN109970996-A (Zhong et al., 2019) demonstrates a process wherein steam explosion is used to separate the fibrils. The BC is filtered to remove fermentation broth, submitted to alkaline treatment, and then chopped and sonicated, resulting in a transparent compound destined for packaging and biomaterial production. The document CN110760091-A (Yang et al., 2020) also show a process that breaks the internal linkages of BC through flash explosion at 200°C to 300°C and 2 to 5 MPa for 10 min. Similarly, patent CN108948383-A (Li et al., 2019) uses liquid nitrogen to solidify the gel and then grinding for 30 min to obtain an ultrafine powder. The BC hydrogels thus obtained can also be applied in the human health field, being mostly destined for biomaterial production. Both examples show how the downstream processing can drastically change the BC properties and final product application.

5.6 CONCLUSIONS

The BC biopolymer has interesting mechanical characteristics to be applied in several products, especially for food and medical purposes. However, much of these features are directly affected by the purification method chosen. The general downstream process for BC involves removing the biofilm or the polymer beads from the fermentation broth; performing a decontamination step to remove cells, toxins, and residual components; and drying. The majority of those operations can be performed through simple equipment and methods, such as centrifugation, filtration, alkaline treatment, lyophilizing, and oven drying. Although all steps can affect or modulate the product properties, the most critical ones, as well as the ones that need further development, are:

- The chemical treatment and drying, which affect porosity, water retention, tensile strength, tensile modulus, and others. The drying step also requires studies even for classical operations, since little information about conditions, parameters, and resulting characteristics are available.

- Decontamination directly affects the material biocompatibility and applications, and still requires research to find alternative chemicals capable of removing cells and LPS properly with efficiency and speed.
- Most of the reported research is on a very small scale. Studies in larger-scale downstream processes for BC are also essential for developing an industrial-scale production of the biopolymer.

Additionally, different strategies are being applied for a great variety of products and properties of BC. In conclusion, BC is an extremely versatile biopolymer that can be applied in several products, and its production is still in development. The downstream plays an important role in this process, being crucial for the resulting mechanical properties, and also requiring wide investigation to optimize this technology.

5.7 REFERENCES

Al-Shamary, Elham Esmaeel, and Amir Khalaf Al-Darwash. 2013. "Influence of Fermentation Condition and Alkali Treatment on the Porosity and Thickness of Bacterial Cellulose Membranes." *The Online Journal of Science and Technology* 3 (2): 194–203.

Amin, Mohd Cairul Iqbal Mohd, Abadi Gumah Abadi, and Haliza Katas. 2014. "Purification, Characterization and Comparative Studies of Spray-Dried Bacterial Cellulose Microparticles." *Carbohydrate Polymers* 99. Elsevier Ltd.: 180–189. doi:10.1016/j.carbpol.2013.08.041.

Amorim, Júlia D. P., Helenise A. Nascimento, Cláudio José G. Silva Junior, Alexandre D. M. Medeiros, Ivo Diego L. Silva, Andréa Fernanda S. Costa, Glória Maria Vinhas, and Leonie A. Sarubbo. 2022. "Obtainment of Bacterial Cellulose with Added Propolis Extract for Cosmetic Applications." *Polymer Engineering & Science* 62 (2). John Wiley & Sons, Ltd.: 565–575. doi:10.1002/PEN.25868.

Ávila, Héctor Martínez, Silke Schwarz, Eva Maria Feldmann, Athanasios Mantas, Achim Von Bomhard, Paul Gatenholm, and Nicole Rotter. 2014. "Biocompatibility Evaluation of Densified Bacterial Nanocellulose Hydrogel as an Implant Material for Auricular Cartilage Regeneration." *Applied Microbiology and Biotechnology* 98 (17). Springer Verlag: 7423–7435. doi:10.1007/S00253-014-5819-Z/FIGURES/8.

Azeredo, Henriette M. C., Hernane Barud, Cristiane S. Farinas, Vanessa M. Vasconcellos, and Amanda M. Claro. 2019. "Bacterial Cellulose as a Raw Material for Food and Food Packaging Applications." *Frontiers in Sustainable Food Systems* 3 (February). Frontiers Media S.A.: 7. doi:10.3389/FSUFS.2019.00007/BIBTEX.

Azmi, Siti Nur Nadhirah Said, Siti Nur Najwa Farhah Mohd Fabli, Fathin Amila Faisul Aris, Zainatul Asyiqin Samsu, Ahmad Syafiq Fauzan Mohd Asnawi, Yuhanees Mohamed Yusof, Hidayah Ariffin, and Sharifah Soplah Syed Abdullah. 2021. "Fresh Oil Palm Frond Juice as a Novel and Alternative Fermentation Medium for Bacterial Cellulose Production." *Materials Today: Proceedings* 42 (January). Elsevier: 101–106. doi:10.1016/J.MATPR.2020.10.220.

Badshah, Munair, Hanif Ullah, Abdur Rahman Khan, Shaukat Khan, Joong Kon Park, and Taous Khan. 2018. "Surface Modification and Evaluation of Bacterial Cellulose for Drug Delivery." *International Journal of Biological Macromolecules* 113 (July). Elsevier: 526–533. doi:10.1016/J.IJBIOMAC.2018.02.135.

Barud, Hernane S., Thaís Regiani, Rodrigo F. C. Marques, Wilton R. Lustri, Younes Messaddeq, and Sidney J. L. Ribeiro. 2011. "Antimicrobial Bacterial Cellulose-Silver Nanoparticles Composite Membranes." *Journal of Nanomaterials*: 1–8. doi:10.1155/2011/721631.

Biospace. 2022. "Microbial and Bacterial Cellulose Production Market To Reach USD 980.9 Million By 2028, Says Reports And Data." January. www.biospace.com/article/microbial-and-bacterial-cellulose-production-market-to-reach-usd-980-9-million-by-2028-says-reports-and-data/.

Cacicedo, Maximiliano L., M. Cristina Castro, Ioannis Servetas, Loulouda Bosnea, Konstantina Boura, Panagiota Tsafrakidou, Agapi Dima, Antonia Terpou, Athanasios Koutinas, and Guillermo R. Castro. 2016. "Progress in Bacterial Cellulose Matrices for Biotechnological Applications." *Bioresource Technology* 213. Elsevier Ltd.: 172–180. doi:10.1016/j.biortech.2016.02.071.

Castro, Cristina, Robin Zuluaga, Jean-luc Putaux, Gloria Caro, Inaki Mondragon, and Piedad Ganan. 2011. "Structural Characterization of Bacterial Cellulose Produced by *Gluconacetobacter swingsii* sp. from Colombian Agroindustrial Wastes." *Carbohydrate Polymers* 84: 96–102. doi:10.1016/j.carbpol.2010.10.072.

Chawla, Prashant R., Ishwar B. Bajaj, Shrikant A. Survase, and Rekha S. Singhal. 2009. "Microbial Cellulose: Fermentative Production and Applications." *Food Technology and Biotechnology* 47 (2): 107–124.

Ebrahimi, Ehsan, Valiollah Babaeipour, and Sirwan Khanchezar. 2016. "Effect of Down-Stream Processing Parameters on the Mechanical Properties of Bacterial Cellulose." *Iran Polymer Journal* 25: 739–746. doi:10.1007/s13726-016-0462-4.

Eichhorn, S. J., A. Dufresne, M. Aranguren, N. E. Marcovich, J. R. Capadona, S. J. Rowan, C. Weder, et al. 2010. "Review: Current International Research into Cellulose Nanofibres and Nanocomposites." *Journal of Materials Science* 45 (1): 1–33. http://doc.rero.ch.

Esa, Faezah, Siti Masrinda Tasirin, and Norliza Abd Rahman. 2014. "Overview of Bacterial Cellulose Production and Application." *Agriculture and Agricultural Science Procedia* 2. Elsevier Srl: 113–119. doi:10.1016/j.aaspro.2014.11.017.

Forte, Ana, Fernando Dourado, Andrė Mota, Belmira Neto, Miguel Gama, and Eugėnio Campos Ferreira. 2021. "Life Cycle Assessment of Bacterial Cellulose Production." *LCA for Agriculture Practices and Biobased Industrial Products* 26: 864–878. doi:10.1007/s11367-021-01904-2.

Gea, Saharman, Christopher T. Reynolds, Nima Roohpour, Basuki Wirjosentono, Nattakan Soykeabkaew, Emiliano Bilotti, and Ton Peijs. 2011. "Investigation into the Structural, Morphological, Mechanical and Thermal Behaviour of Bacterial Cellulose after a Two-Step Purification Process." *Bioresource Technology* 102 (19). Elsevier Ltd: 9105–9110. doi:10.1016/j.biortech.2011.04.077.

Gromovykh, T. I., N. B. Feldman, O. A. Tikhonova, S. V. Lutsenko, P. S. Timashev, K. N. Bardakova, S. N. Churbanov, O. I. Kiselyova, M. N. Kraeva, and A. S. Grinevich. 2018. "Elaboration of a Bacterial Cellulose Matrix for the Immobilisation of *Escherichia coli* Cells." *International Journal of Nanotechnology* 15 (4–5). Inderscience Publishers: 288–300. doi:10.1504/IJNT.2018.094784.

Hu, Yang, Jeffrey Catchmark, and Erwin Vogler. 2013. "Factors Impacting the Formation of Sphere-like Bacterial Cellulose Particles and Their Biocompatibility for Human Osteoblast Growth." *Biomacromolecules* 14: 3444–3452. doi:10.1021/bm400744a.

Krystynowicz, A., W. Czaja, A. Wiktorowska-Jezierska, M. Gonçalves-Miśkiewicz, M. Turkiewicz, and S. Bielecki. 2002. "Factors Affecting the Yield and Properties of Bacterial Cellulose." *Journal of Industrial Microbiology and Biotechnology* 29 (4): 189–195. doi:10.1038/sj.jim.7000303.

Li, Z., C. Dou, Y. Liu, J. Gong, and J. Zhang. 2019. "Preparation of Bacterial Cellulose Ultrafine Powder Used for Preparing Composite Hydrogel by Purifying Bacterial Cellulose Wet Film, Cutting, Adding to Liquid Nitrogen, and Grinding." CN108948383-A, issued 2019.

Liebner, Falk, Emmerich Haimer, Martin Wendland, Marie Alexandra Neouze, Kerstin Schlufter, Peter Miethe, Thomas Heinze, Antje Potthast, and Thomas Rosenau. 2010. "Aerogels from Unaltered Bacterial Cellulose: Application of ScCO2 Drying for the Preparation of Shaped, Ultra-Lightweight Cellulosic Aerogels." *Macromolecular Bioscience* 10 (4): 349–352. doi:10.1002/mabi.200900371.

Liu, J., Q. Zhang, Y. Cai, J. Zhang, and H. Su. 2022. "Method for Purifying Bacterial Cellulose Membrane Useful in Preparation of Artificial Skin, External Dressing, Artificial Blood Vessel, Cartilage Tissue, Vibrating Membrane and Slow Release Carrier Comprises e.g. Washing Bacterial Cellulose Membrane With." CN113881085-A, issued 2022.

Liu, X., and L. Shi. 2018. "Purifying Bacterial Cellulose Film Involves Screening Bacterial Cellulose Film Material, Pretreating Bacterial Cellulose Film Material with Water, and Removing Pretreated Bacterial Cellulose Film Cell Endotoxin. CN107582509-A, issued 2018.

Masoodi, R., R. F. El-Hajjar, K. M. Pillai, and R. Sabo. 2012. "Mechanical Characterization of Cellulose Nanofiber and Bio-Based Epoxy Composite." *Materials and Design* 36. Elsevier Ltd: 570–576. doi:10.1016/j.matdes.2011.11.042.

Matsuoka, Masanobu, Takayasu Tsuchida, Kazunobu Matsushita, Osao Adachi, and Fumihiro Yoshinga. 1996. "A Synthetic Medium for Bacterial Cellulose Production by *Acetobacterxylinum* subsp. *sucrofermentans*." *Bioscience, Biotechnology, and Biochemistry* 60 (4): 575–579.

Mikkelsen, D., B. M. Flanagan, G. A. Dykes, and M. J. Gidley. 2009. "Influence of Different Carbon Sources on Bacterial Cellulose Production by Gluconacetobacter xylinus Strain ATCC 53524." *Journal of Applied Microbiology* 107 (2). John Wiley & Sons, Ltd: 576–583. doi:10.1111/J.1365-2672.2009.04226.X.

Moniri, Mona, Amin Boroumand Moghaddam, Susan Azizi, Raha Abdul Rahim, Arbakariya Bin Ariff, Wan Zuhainis Saad, Mohammad Navaderi, and Rosfarizan Mohamad. 2017. "Production and Status of Bacterial Cellulose in Biomedical Engineering." *Nanomaterials* 7: 1–26. doi:10.3390/nano7090257.

Moon, Robert J., Ashlie Martini, John Nairn, John Simonsen, and Jeff Youngblood. 2011. "Cellulose Nanomaterials Review: Structure, Properties and Nanocomposites." *Chemical Society Reviews*. 40. doi:10.1039/c0cs00108b.

Peng, Yucheng, Douglas J. Gardner, and Yousoo Han. 2012. "Drying Cellulose Nanofibrils: In Search of a Suitable Method." *Cellulose* 19 (1): 91–102. doi:10.1007/s10570-011-9630-z.

Pigaleva, Marina A., Matvey V. Bulat, Tatiana I. Gromovykh, Irina A. Gavryushina, Sergey V. Lutsenko, Marat O. Gallyamov, Ilya V. Novikov, Anastasiia G. Buyanovskaya, and Olga I. Kiselyova. 2019. "A New Approach to Purification of Bacterial Cellulose Membranes: What Happens to Bacteria in Supercritical Media?" *Journal of Supercritical Fluids* 147 (January): 59–69. doi:10.1016/j.supflu.2019.02.009.

Procentese, Alessandra, Francesca Raganati, Giuseppe Olivieri, Maria Elena Russo, and Antonio Marzocchella. 2017. "Pre-Treatment and Enzymatic Hydrolysis of Lettuce Residues as Feedstock for Bio-Butanol Production." *Biomass and Bioenergy* 96 (January). Pergamon: 172–179. doi:10.1016/J.BIOMBIOE.2016.11.015.

Putra, Ananda, Akira Kakugo, Hidemitsu Furukawa, Jian P. Gong, and Yoshihito Osada. 2008. "Tubular Bacterial Cellulose Gel with Oriented Fibrils on the Curved Surface." *Polymer* 49 (7): 1885–1891. doi:10.1016/j.polymer.2008.02.022.

Reiniati, Isabela, Andrew N. Hrymak, and Argyrios Margaritis. 2017. "Kinetics of Cell Growth and Crystalline Nanocellulose Production by *Komagataeibacter xylinus*." *Biochemical Engineering Journal* 127 (November). Elsevier: 21–31. doi:10.1016/J.BEJ.2017.07.007.

Saska, Sybele, Lucas Novaes Teixeira, Paulo Tambasco De Oliveira, Ana Maria Minarelli Gaspar, Sidney José Lima Ribeiro, Younès Messaddeq, and Reinaldo Marchetto. 2012. "Bacterial Cellulose-Collagen Nanocomposite for Bone Tissue Engineering." *Journal of Materials Chemistry* 22 (41). Royal Society of Chemistry: 22102–22112. doi:10.1039/C2JM33762B.

Skiba, Ekaterina A., Vera V. Budaeva, Elena V. Ovchinnikova, Evgenia K. Gladysheva, Ekaterina I. Kashcheyeva, Igor N. Pavlov, and Gennady V. Sakovich. 2020. "A Technology for Pilot Production of Bacterial Cellulose from Oat Hulls." *Chemical Engineering Journal* 383 (March). Elsevier: 123128. doi:10.1016/J.CEJ.2019.123128.

Vasconcellos, Vanessa M., and Cristiane S. Farinas. 2018. "The Effect of the Drying Process on the Properties of Bacterial Cellulose Films from *Gluconacetobacter hansenii*." *Chemical Engineering Transactions* 64: 145–150. doi:10.3303/CET1864025.

Wang, Jing, Javad Tavakoli, and Youhong Tang. 2019. "Bacterial Cellulose Production, Properties and Applications with Different Culture Methods: A Review." *Carbohydrate Polymers* 219. Elsevier: 63–76. doi:10.1016/j.carbpol.2019.05.008.

Yang, J., W. Li, and X. Li. 2020. "Preparing Modified Bacterial Cellulose Whisker by Hydrolyzing Purified Bacterial Cellulose Floc, and Treating Bacterial Cellulose Whiskers Using Flash Explosion Technology, and Drying. CN110760091-A, issued 2020.

Zeng, Muling, Anna Laromaine, and Anna Roig. 2014. "Bacterial Cellulose Films: Influence of Bacterial Strain and Drying Route on Film Properties." *Cellulose* 21 (6): 4455–4469. doi:10.1007/s10570-014-0408-y.

Zhang, B., M. Zhang, S. Yang, and Y. Wu. 2022. "Preparing Amorphous Silver-Containing Long-Acting Antibacterial Hydrogel Dressing for Wounds in Different parts, by Washing Bacterial Cellulose Membrane with Deionized Water and Purifying, Hydrolyzing and Dialyzing, Adding Tris Buffer, Stirring, Washing a." CN113980185-A, issued 2022.

Zhong, C., B. Li, S. Jia, X. Hu, and Y. Xie. 2019." Dissolving Bacterial Cellulose for Biomedical Materials, Involves Purifying, Chopping, Homogenizing, Sonicating and Filtering Bacterial Cellulose, Treating Filtered Bacterial Cellulose by Steam Explosion Technology." CN109970996-A, issued 2019.

6 Characterization and Storage of Bacterial Cellulose

Aida Fadakar Sarkandi and Majid Montazer
Amirkabir University of Technology
Tehran, Iran

6.1 INTRODUCTION

Bacterial cellulose (BC), which is inherently a hydrogel, has a three-dimensional nanofibrous network with special properties, including excellent mechanical properties, porosity, and high water absorption and retention (Gao et al., 2011). These properties depend on various factors, such as the type of bacteria used for production, condition and type of culture medium, temperature, pH, neutralization and purification, and the method of drying (Krystynowicz et al., 2002; Angela, 2019). The appropriate drying method is chosen according to the final application, properties, cost, and environmental effects (Illa et al., 2019). In this chapter, an attempt is made to provide a suitable background to help determine the appropriate technique and method for investigation and also the final application of BC.

Here, the best and most practical techniques that are mainly used to investigate BC, as well as their limitations, are discussed. In each section, the related techniques that can be used for characterization are thoroughly described. to the aim is for researchers to learn about the reason for using these specific techniques. The standard techniques and methods for sample characterizing are also provided. In addition, the drying conditions and their effect on the microstructures and other properties of bacterial cellulose are investigated and compared.

An important point to mention is that researchers should be aware of the level of characteristics required for their needs to avoid the unnecessary application of all techniques that are contrary to the purpose of this chapter.

6.2 CHARACTERIZATION OF BACTERIAL CELLULOSE

6.2.1 Elemental Analysis of Bacterial Cellulose

6.2.1.1 Carbon, Hydrogen, Nitrogen Elemental Analysis

Carbon, hydrogen, nitrogen elemental analysis (CHN EA) is used to determine the elements of organic substrates based on the Dumas method that pure oxygen reacts with all the available carbons in the sample at a very high temperature (above 1000°C) and turns it into carbon dioxide. In addition, oxygen reacts with available hydrogen and nitrogen in the sample and turns into water and nitrogen oxide, respectively. With the progress in this technology, it is possible to identify more chemicals, including oxygen utilizing carbon and copper oxide (CHNS-O), sulfur using barium sulfate (CHNS), and halogens by silver nitrate (CHNSX). To detect these elements, a new detector or an addition to the device is needed (Steyermark 2012; Foster et al., 2018).

After the first reaction step, all chemical substances in the sample except hydrogen (H), oxygen (O), carbon (C), nitrogen (N), and sulfur (S) are converted into combustible substances, which are absorbed by various absorbents and prevent external contamination. The desired combustion products are removed from the combustion chamber using inert helium gas (Foster et al., 2018). The gas

 DOI: 10.1201/9781003355434-6

flow is over high-purity copper, which is used to remove unreacted oxygen and also to convert nitrogen oxide into nitrogen gas. Gas chromatography, infrared thermal conductivity cells, and chemical traps are among the gas detection methods, depending on the detector used in the device (Foster et al., 2018).

In general, CHN EA is used in three basic applications: proving acid hydrolysis by sulfuric acid, functionalizing materials based on nanocellulose with a functional group or a molecule, and measuring the degree of chemical leaching of polymers containing nanocellulose. Among these methods, acid hydrolysis is particularly useful because the remaining sulfur groups from acid hydrolysis can be easily identified in the output spectrum. The second application, the functionalization of cellulose-based materials, is a simple method if the functional groups contain nitrogen and sulfur, due to the very easy identification of these groups in the output spectra. Furthermore, if the groups contain carbon, oxygen, and hydrogen, the elements present in the nanocellulose itself, the output spectrum will show an increase in the peak for the mentioned elements. Finally, the important point in the third application is the impossibility to perform CHN EA on a liquid sample and the samples must be dried (Neto Flauzino et al., 2016; Pang et al., 2010; Tonoli et al., 2012; Takegawa et al., 2009).

CHN EA also has its limitations and the acceptable range for the samples containing carbon and nitrogen is about 500 ppm, meanwhile, some others reported lower values around 200–300 ppm. Therefore, it is necessary to be careful in using CHN EA in low and insignificant loadings. Moreover, due to chemical limitations, this method cannot be used to detect elements other than C, H, N, O, S, and halogens in cellulosic samples. Also, due to the presence of C, O, and N as the main structural elements of cellulose, it is very difficult to distinguish between these elements before and after chemical modifications (Foster et al., 2018; Carbon, 2011).

6.2.1.2 Energy-Dispersive X-Ray Spectroscopy

Energy-dispersive x-ray spectroscopy (EDS) is an analytical technique used for sample analysis. The ground state electrons in the atoms of the sample are excited by an external source and after leaving their shells, an electron hole is created. Therefore, an electron from the outer layer can fill the created hole and produce an x-ray. The number and energy of these x-rays can be measured using an energy-dispersive spectrometer and used to analyze the samples. Because the energy-dispersive spectrometer can be attached to most scanning electron microscopes (SEMs), it is easy to obtain the chemical composition of surfaces that can be seen in the SEM (McHenry and Laughlin, 2012) In some studies, EDS was used to analyze silver nanoparticles on bacterial cellulose (Fadakar Sarkandi et al., 2021) determine the presence of phosphorous and calcium in scaffolds prepared from bacterial nanocellulose (Paramasivan et al., 2022), and analyze bacterial cellulose-cement composite Mohammadkazemi et al., 2015.

Specifying light elements and the lack of sensitivity required for chemical tracing are among the limitations of EDS; therefore, it cannot be used to determine a negligible quantity of atoms (Foster et al., 2018; Kumar et al., 2014).

6.2.1.3 X-Ray Photoelectron Spectroscopy

X-Ray photoelectron spectroscopy (XPS), or electron spectrometry, is one of the methods of chemical analysis. When the surface of the material is irradiated with x-rays, the electrons of the surface atoms are energized by the x-rays and are converted into photoelectrons. Finally, the energy of the emitted ions is used to analyze the composition of the materials. XPS is used to analyze carbon bonds in chemical modifications and also to identify heavy ions in compounds containing nanocellulose (Briggs, 1977; Siegbahn, 1982).

XPS, similar to the other methods, has limitations such as the solidity of the sample, and the lack of sensitivity to identify chemical substances. In addition, another main point is carbon contamination, which usually has a significant interference with the output spectra of cellulosic samples (Johansson and Campbell, 2004; Reid et al., 2017).

6.2.1.4 Secondary Ion Mass Spectrometry

In secondary ion mass spectrometry (SIMS) analysis, the surface of a material is determined by bombarding it using an ion beam such as argon ions (Ar^+), gallium ions (Ga^+), or an alkali metal ion. In short, these ions cause secondary ions to leave the surface of the material and enter the atmosphere near the sample, and then these secondary ions are filtered by a mass spectrometer and analyzed (Griffiths, 2008).

There are different modes for SIMS analysis. The first is the static SIMS, in which the surface is sputtered in the same place and can usually be used to determine the intermediate stages of a catalyzed reaction and also to check the number of material changes with depth. In this case, the detection depth of the material is 1.5 nm. The second is the scanning SIMS. In this method, the surface of the material is scattered to a short depth, but a wide area is scanned. This is used to identify the differences in chemical concentration on the surface of the sample, and to determine the presence or absence of surface pollution and the structure of thin layers. The final method is the dynamic mode of SIMS, which uses a high-intensity beam to cut the sample using the quick sputtering technique. It is possible to measure compounds as a function of depth; however, the sample is permanently damaged (Griffiths, 2008; Sodhi, 2004; Gritsch et al., 1999).

6.2.1.5 Inductively Coupled Plasma-Mass Spectrometry

The basis of inductively coupled plasma-mass spectrometry (ICP) is the plasma used to ionize the sample atoms. In this method, to light the torch, first argon gas is sent up through the central pipe then the spark created by the induction coil ionizes the gas. The ions and electrons resulting from ionization react with the magnetic field produced by the induction coil, causing the flow of electrons and ions in specific circular paths in the system. The studied sample in ICP, after evaporation under the influence of electron energy and environmental ions, is transformed into its constituent atoms and excited in a very hot plasma environment. Finally, these ions are introduced into a spectrometer using a series of cones and are separated based on the mass-to-charge ratio (Huang and Hieftje, 1989; Hill, 2008).

The detection of metal compounds, salts, and impurities in the sample is among the main applications of this method. For instance, the ICP method was used to identify palladium leaching in palladium nanoparticles synthesized on nanocellulose (Cirtiu et al., 2011). The lack of applicability in the analysis of light elements, as well as the limitation of conducting tests on liquid samples, are among the limitations of this method (Foster et al., 2018).

6.2.2 Electron Microscopy

In electron microscopy, electrons are used instead of photons because electron beams are accelerated under high voltage with a shorter wavelength, making the output information in the nanometers range. In general, there are various types of microscopes used in electron microscopy, including scanning electron microscope (SEM), transmission electron microscope (TEM), and field emission scanning electron microscope (FESEM). Each of these are explained separately in the following.

6.2.2.1 Scanning Electron Microscopy

The SEM is one of the most widely used electron microscopes in the world, invented in 1935 by Max Knolls in an experiment in Germany and then commercially launched in 1965 (Zhou and Wang, 2007). Before explaining scanning electron microscopy, a short section about light microscopy is presented, which will be of great help in understanding the operation of electron microscopy.

In optical microscopy, visible light (electromagnetic waves in the visible region) is used as an image-generating factor. In this way, after passing through one or more lenses, the visible light is focused, then hits the sample and is reflected. Due to the refraction of the light in the collision with the sample and the reflection of the light rays in different directions, the image is formed on the eyepiece of the microscope or a monitor. In other words, the mechanism of image formation is the refraction of light. If the light rays are not refracted, a completely uniform image (a white screen) will be obtained. The length of the wavelength of visible light is 400 to 700 nm, thus particles smaller than this range cannot be seen with visible light, even at very high magnifications (Davidson and Abramowitz, 2002).

In scanning electron microscopy, instead of visible light, an electron beam is used as the image-generating factor, which greatly increases the magnification and resolution; then particles as small as a few nanometers will be visible. Due to the impact of the electron beam on the studied sample, various phenomena occur. Knowledge of these phenomena is necessary for a precise understanding of the image formation mechanism in the SEM (Zhou and Wang, 2007; Mayeen et al., 2018). A description of the structure of the SEM will make the mechanism of image formation clear.

The SEM has three main parts: an electron gun, an electron column, and a device housing. The role of the electron gun is to produce electrons continuously and stably, as well as to form an electron beam. In addition, another task of the electron gun is to accelerate the electrons toward the sample. After the electron beam is produced and directed, the diameter of the electron beam should be reduced to increase the number of electrons interacting with the material. This is done by the electron column; two electromagnetic lenses are used that focus the electron beam, similar to optical lenses in an optical microscope, with the difference that the focal distance of these lenses is not fixed. For better device performance, the SEM must work in a vacuum; therefore, the sample is placed in the chamber of the device, the vacuum system is activated, and after reaching the desired pressure, imaging begins (Zhou and Wang, 2007).

The important advantage of scanning electron microscopy is to provide a direct and point-to-point image of the sample surface with high resolution. Other advantages include not requiring a transparent sample, checking all conductive and nonconductive samples, the possibility of imaging in three dimensions, easy initial preparation of the samples, ease of working with the device, and high speed. Despite its many advantages, it also has limitations, including the need for vacuum and generating only black-and-white images (Zhou and Wang, 2007; Afanasyev et al., 2019).

6.2.2.2 Transmission Electron Microscopy

The TEM is one of the most powerful tools available to study materials and their structure at the nanoscale with a resolution of 0.1 to 1 nm and a magnification of 1,500,000. At these magnifications a particle with dimensions in angstroms can be seen. It is one of the special tools for determining the structure and morphology of materials that can be used to study structure, symmetry, orientation, and crystal defects. Like the SEM, electrons are used instead of light. The TEM includes an electron gun, lenses, sample chamber, image system, screen, and vacuum pump. The operation method of the TEM is briefly described as follows: an electron beam is passed through the surface of the sample and after passing through the lenses it hits the fluorescent screen and forms an image. The color of these images depends on the number of electrons passing through the sample. Brighter areas have received more electrons and darker areas have received fewer electrons. There are two types of images, light background and dark background (Carter and Williams, 2016; Zuo and Spence, 2017; Parker, 2001).

In bright background images, in the areas where the sample thickness is greater, the image is seen as dark and the background as light. These types of images are used to investigate crystal states, conditions, and grain defects due to the good representation of the microstructure. On the other hand, in the dark background image, the background image is dark and in the sample areas, the image is light (Zuo and Spence, 2017; Parker, 2001).

The limitations of transmission electro microscopy include the difficult and time-consuming process, low resolution (about 0.2 nm), small field of view, and radiation damage to the sample. Also, these microscopes are not able to create color images and cannot analyze gaseous, liquid, and thick samples. In the TEM, the thickness of the sample must be very thin because a large number of electrons must pass through it. In sample preparation, the sample must first be made thin. This operation is performed using mechanical methods such as mechanical polishing or chemical methods such as electropolishing. Nonconductive samples should also be made conductive by using a thin layer of gold or graphite (Carter and Williams, 2016; Zuo and Spence, 2017).

6.2.2.3 Comparing Scanning Electron Microscopy with Transmission Electron Microscopy

In the SEM, the beam hits the surface of the material and analyzes the surface of the sample, whereas in the TEM, the beam passes through the sample and examines the inside of the sample. For this reason, the acceleration of electrons in a TEM is higher than in a SEM. As a result, sample preparation in a TEM is more difficult than in a SEM because the sample in these microscopes must be thin. Another important difference between these two microscopes is the resolution and magnification. The magnification and resolution in the TEM can be up to 0.1 nm and 50 to 1,500,000 times, but in the SEM are 0.5 nm and 10 to 1,000,000 times. In addition, sample analysis in the SEM is line by line, while the TEM has a wide electron beam and analyzes the entire sample in one step (Foster et al., 2018; Akhtar et al., 2018).

6.2.2.4 Field Emission Scanning Electron Microscope

A scanning electron microscope in which primary electrons are produced by a field emission electron gun is called a field emission scanning electron microscope (FESEM). In the SEM, a thermal emission electron gun is used, which is usually made of tungsten filaments. Electric currents pass through these filaments and lead to the creation of heat around 2800 K.

The only major difference between the FESEM and the SEM is the electron gun and the production of primary electrons. In the field emission electron gun, similar to thermal emission, there is a tungsten filament with a very sharp tip of about 50 nm, but instead of heating to form primary electrons, a very strong electric field is used to separate electrons from the surface of the tungsten filament. In the FESEM, the primary electrons have a higher speed to reach the surface of the sample due to the created electric field, which reduces the wavelength of the electrons. It leads to creating an image with a better resolution than the SEM (Mayeen et al., 2018; Akhtar et al., 2018; Dey et al., 2020).

6.2.2.5 Challenges of Electron Microscopy

The SEM and TEM provide useful information to researchers through the identification of materials, although they present some challenges. All materials, especially bacterial cellulose, have unique characteristics that require preparation protocols for the examination of samples and appropriate imaging parameters (Foster et al., 2018). In this part of the chapter, the challenges of bacterial cellulose imaging using electron microscopy are discussed.

Among the challenges of imaging BC through electron microscopes are low thickness and low electron density, which cause difficulty. The BC nanofiber has a low thickness and elements with low atomic numbers, which causes less signal to be produced and, hence, the signals close to the background noise are not seen in the image. The signal is the number of electrons that reach the detector of electron microscopy. This challenge is present in both SEM and TEM microscopes, which leads to a low difference between the bright and dark parts of the final image or low image contrast. Therefore, creating contrast during sample preparation is very important (Buckley et al., 1993; Özdöl et al., 2012). Another key point is the sample thickness and its significant effect on image resolution. For instance, in TEM, a sample with a thickness of 1 mm or less is needed for the movement of the electron beam. Only high-resolution SEM (or FESEM) can image bacterial nanocellulose because the diameter of the fibers is usually below 100 nm and it is difficult to image them (Özdöl et al., 2012).

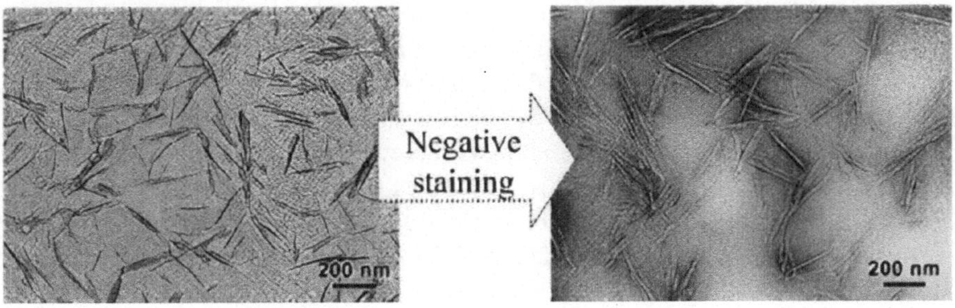

FIGURE 6.1 Transmission electron microscope images of cellulose nanocrystals (a) with negative staining and (b) unstained (Shojaeiarani et al., 2021).

There are various methods to improve the quality of electron microscopy images, and each of these can be used individually or in combination. It is necessary to conduct some tests to find the appropriate general method because the best method depends on the constituents of the sample, size, surface function, and other variables. Here are some tips on how to improve contrast.

In transmission electron microscopy, the basic way to reduce the contrast problem is by using heavy metal elements through staining, while in scanning electron microscopy, the key way is to cover the sample with gold or platinum. These coatings make the nanocellulose stand out, and as a result, the final image will be better. Staining is done after placing the samples on a suitable substrate. In general, there are two types of positive and negative staining, among which negative staining is more common due to its ease of use. In negative staining, the outline of the sample is surrounded and causes the background of the sample to have more contrast, while in positive staining, the spots are chemically bonded to the sample and lead to an increase in the contrast (Foster et al., 2018; Ayache et al., 2010; Kaushik et al., 2015). The difference between the TEM images of stained and unstained samples is shown in Figure 6.1.

Moreover, another fundamental challenge of electron microscopes is charging and sample damage. Since the mechanism of electron microscopy is based on directing electrons toward the sample and then causing the interaction and exit of electrons, it leads to sample damage in nonconductive samples. In other words, in conductive samples, electron exchange is done with the least change in the structure of the material, such as metals. However, in bacterial nanocellulose, due to the low number of electrons and its nonconductivity, loading occurs and, as a result, the sample is destroyed. Damage in the BC sample under electron beams occurs due to the generation of high voltage in a very small area by the electron beam, which destroys organic materials, especially if the beam stays in one area for a long time (Ayache et al., 2010; Michen et al., 2015). To reduce the charge and destruction of the sample, it is recommended to use gold and platinum coating in scanning electron microscopy, as well as the staining method in transmission electron microscopy (Foster et al., 2018; Buckley et al., 1992).

6.2.3 Atomic Force Microscopy

The atomic force microcope (AFM) was built in 1986 at NEC company by scientists Gerd Binnig and Quate to study the mechanical properties of surfaces. The AFM processes an image of the material surface by measuring interatomic forces at the nanoscale. The measurement of these forces is based on the formula (1), in such a way that if a force (F) is applied to the spring, the spring changes length by the amount X.

$$F = -K.X \tag{1}$$

in which K is the spring constant that resists change in the spring length. If a nanometer spring is moved in a straight line perpendicular to the surface of a solid, the spring is compressed in surface protrusions and stretched in depressions. If the amount of change in the length of the spring during the movement of the spring on the surface can be measured, it is possible to draw the protrusions and depressions of the surface of the material in its survey line (Bellitto, 2012; Haugstad, 2012).

In the AFM, the surface of the material is sensed by a sharp needle or tip. The needle is located on a base called a cantilever with a length of about 100 to 400 μm. The cantilever must have high elasticity to show mechanical movement due to electrical stimulation, thus it is usually made of silicon or silicon nitride having high resistance to erosion.

The needle is placed at a distance of a few angstroms from the surface of the sample. It is very sharp in a manner that ideally only one atom can fit in its tip. The atoms at the tip of the needle and the surface atoms of the material exert forces on each other leading to the curvature of the cantilever. The curvature of the cantilever and the force acting on it, which behaves as a spring, are recorded with high accuracy. Using a laser that analyzes the movements of the cantilever, a signal is sent to the detector and recorded. For more laser reflection, the cantilever is covered with very thin layers of metals such as gold, aluminum, or silver. By scanning the entire surface of the sample line by line, a three-dimensional image of the surface or surface topography is obtained and the surface properties such as depth and height, adhesion, friction, and magnetic properties are obtained (Haugstad, 2012; Eaton and west, 2010). AFM images of bacterial cellulose are shown in Figure 6.2.

The forces between the needle and the sample surface can be van der Waals, electrostatic, capillary, and magnetic forces. Depending on these forces and the distance between the needle and the surface of the sample, the imaging modes are divided into three categories: contact, noncontact, and tapping (Haugstad, 2012).

The mode used for the analysis of hard and soft samples is different. The contact mode is used for the analysis of hard samples and the noncontact method is used for soft and biological samples. In addition, the tapping mode is used to examine vulnerable samples when high resolution is required. The resolution of the noncontact mode is lower than the contact mode (Bellitto, 2012; Haugstad, 2012; Revin et al., 2018). A comparison of different modes of atomic force microscopy is given in Table 6.1.

Compared to other analysis methods, atomic force microscopy has easier preparation steps and application. Also, the ability to work in both vacuum and air environments and reduced costs distinguishes the use of the AFM from other methods of analysis. Moreover, one of the most important advantages of the AFM is that it can be used for conductive, insulating, and semiconductive samples with no limitations in terms of the physical properties of the materials. The restricted magnification range and the possibility of damage to the needle tip or the sample are the limitations of the AFM (Bellitto, 2012; Haugstad, 2012; Chatterjee et al., 2010).

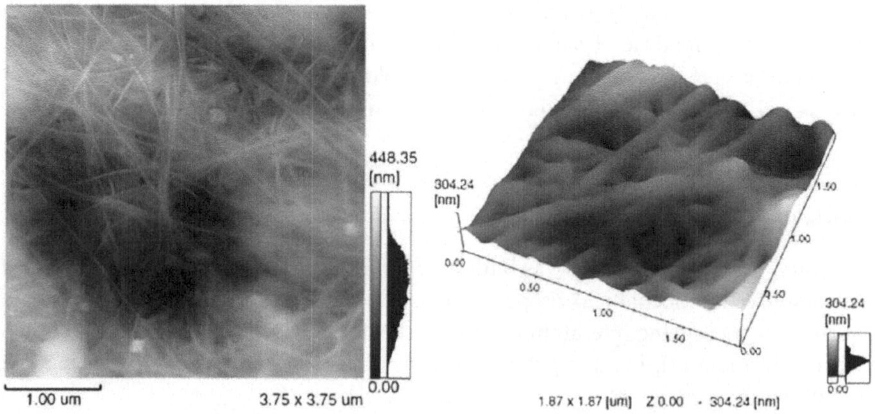

FIGURE 6.2 Atomic force microscopy images of bacterial cellulose (Revin et al., 2018).

TABLE 6.1
Comparison of Various Atomic Force Microscope (AFM) Modes

AFM Mode	Advantages	Disadvantages
Contact	High scanning speed and high resolution, as well as scanning convenience in hard samples	Lower resolution, the negative effect of lateral forces on the image, and also damage to soft samples
Noncontact	No damage to the sample, atomic resolution power in high vacuum, and suitable for soft samples	Low scanning speed
Tapping	High resolution, removal of side forces, and less damage to the sample	Low scanning speed

6.2.4 Surface Modification Characterization

6.2.4.1 Fourier Transform Infrared Spectroscopy Attenuated Total Reflection (FTIR-ATR)

Fourier transform infrared spectroscopy (FTIR) is one of the most widely used methods in the qualitative identification of molecules, used for determining the molecular structure of different species and functional groups in the structure of a substance (Lis et al., 2005). FTIR devices have many advantages, including high speed, high resolution (less than 0.1 cm^{-1}), accurate measurements, good repeatability, good sensitivity, and the absence of unwanted radiation due to the use of an interferometer (Chen et al., 2015).

In FTIR, instead of irradiating a monochromatic beam, a beam with thousands of different frequencies or wavelengths is irradiated to the sample at the same time, and the absorption is analyzed. In the next step, another beam with thousands of other wavelengths is irradiated to the sample to produce the next spot; this is repeated several times. Finally, all this information is collected by the computer and analyzed to detect the absorption of the sample at each wavelength point (Chen et al., 2015).

Attenuated total reflection (ATR) spectra can be obtained with very good quality with FTIR spectrometers. In the ATR technique, the properties of total internal reflection are used. Infrared radiation passes through the ATR crystal reflected at least once from the inner surface in contact with the sample. The depth of penetration of radiation into the sample is usually between 0.5 and 2 μm; the exact amount depends on the wavelength of light, the angle of radiation and the reflection indices for the ATR crystal, and the characteristics of the sample. The ATR technique is generally used in the IR region (approximately 0.8–2.5 μm wavelength) to capture absorption spectra of thin films and opaque materials. The FTIR spectrum can be analyzed by examining different regions (Chen et al., 2015; Elmer, 2005). Table 6.2 can be helpful for the analysis of different regions of the spectrum obtained from bacterial cellulose.

TABLE 6.2
Attenuated Total Reflection-Fourier Transform Infrared Spectroscopy Peaks and the Corresponding Functional Groups for Bacterial Cellulose

	Band (cm^{-1})	Assignments
1	3200–3700	Degree of esterification of the hydroxyl groups
2	3000–3200	Stretching vibrations of CH, CH_2, and CH_3 groups
3	2700–3200	CH groups linked to double bonds or aromatic structures
4	2000–2600	OD and SH groups, C≡N and C≡C triple bonds
5	1600–1800	C=O, C=C, and N=O double bonds and the deformation vibrations of amino groups

6.2.5 Solid-State Properties

6.2.5.1 X-Ray Diffraction Analysis

X-ray diffraction analysis (XRD) can be used to analyze all cellulosic materials. XRD is used to calculate the degree of crystallinity and also to check the factors affecting the ordered and unordered areas (Park et al., 2010). The XRD device is widely used in the identification of unknown crystalline solid samples; thus a powder diffractometer equipped with a CuXa x-ray tube, a detector with 40 kV and 40 mA generator settings, and a graphite monochromatic diffractometer are needed. Data should be recorded in 2θ=5–90° with 0.02° step size and 1 s time in each step.

It is necessary to point out that the angle of the maximum peak indicates the distance between the crystal plates, and the intensity of each peak contains information about the arrangement of atoms. According to Bragg's law, when a ray hits an atom in different layers, the reflected ray of the lower layers has a path different from the layer before, which is defined as follows:

$$\Delta X = 2dsin\theta \tag{2}$$

In this equation, *d* is the distance between crystal planes. The path difference is a multiple of the wavelength (λ) of the ray and is defined as follows:

$$n\lambda = 2dsin\theta \tag{3}$$

In this equation usually, *n* is 1.

The phase difference between the reflected rays of the layers is defined in terms of the path difference as follows:

$$\Delta\phi = 2\pi\Delta x / \lambda \tag{4}$$

If the phase difference of the two reflected rays is an integer multiple of 2π, the two rays create constructive interference and the maximum peak is created. When Bragg's law is established, the average distance between the crystal planes can be obtained by using the information in the x-ray pattern. Also, considering the intensity of maximum peaks and comparing them with standard patterns, the structure of atoms and crystal phase can be determined.

The average size of the particles is calculated using the width of the largest and most symmetrical peak using Scherer's equation.

$$D = k\lambda / \beta cos\theta \tag{5}$$

where β is the peak width at half height, θ is the Bragg angle corresponding to the peak, and λ is the x-ray wavelength (Khan et al., 2020; Nasrollahadeh et al., 2019).

XRD is a low-cost and widely used technique in which the cost of construction is reduced because there is no need for a vacuum. Also, XRD is a noncontact and nondestructive technique, and its preparation is not complicated. The low resolution, as well as the low intensity of the diffracted rays, are among the disadvantages of XRD. The intensity of x-ray radiation depends on the atomic number, and for light elements, this is low, causing problems (Foster et al., 2018; Khan et al., 2020; Nasrollahzadeh et al., 2019).

6.2.5.2 Nuclear Magnetic Resonance Spectroscopy

Nuclear magnetic resonance (NMR) spectroscopy is a method used to determine the magnetic properties of some elements. NMR analysis is based on the nuclear magnetic resonance property and can provide useful information about the structure, dynamics, and state of chemical reactions of molecules. The physical and chemical properties of atoms or molecules can be determined; hence, it is widely used in identifying the structure of organic compounds or the characteristics of organic molecules (Singh and Singh, 2021; Kurukavak and Yilmaz, 2021).

In this test, the sample is placed in a magnetic field and the NMR signal is converted into nuclear magnetic resonance by stimulating the sample nucleus with radio waves, which are detected by sensitive radio receptors. Since the intramolecular magnetic field around an atom in the molecule changes the resonance frequency, it is possible to access the details of the electronic structure of a molecule and its unique functional groups. For NMR spectroscopic testing, the samples must be soluble, and special solvents without protons are used for their preparation, such as deuterated chloroform, which contains deuterium instead of hydrogen as usually used for nonpolar compounds. For polar compounds, other deuterated solvents such as acetone, dimethyl sulfoxide, and water are used (Foster et al., 2018; Zhang et al., 2020).

NMR spectra are unique, analytically tractable, and often very predictable for small molecules. However, the technique also has limitations, such as the relatively large amount of sample required. The time interval of NMR is also relatively long, making it unsuitable for observing fast phenomena, and NMR spectroscopy produces only an average spectrum. In addition, NMR analysis is a very expensive technique and requires special skills to work. It is ineffective for samples with low concentrations, and the maintenance and repair of the device are also very expensive (Elipe, 2003).

6.2.5.3 Raman Spectroscopy

Raman spectroscopy is a nondestructive chemical analysis technique that provides detailed information about chemical structure, phases, polymorphology, crystallinity, and molecular interactions (Smith and Dent, 2019). This is based on the interaction of light with chemical bonds within a substance. Raman spectroscopy has been used since 1970 in the analysis of cellulosic materials and since 2005 to estimate crystallinity (Schenzel et al., 2005).

The functioning of the Raman spectrometer is, first, the laser light hits the beam splitter and is divided into two parts. One part hits the sample and the other part hits the filter, and the laser scatters after hitting the sample. The scattered light is directed to the beam splitter by objective lenses, and then the whole light passes through it and hits the filter. The laser wavelength is filtered by this element, but other wavelengths pass through this filter. The filtered light is focused by the slit and then by a lens on the mirror. Then, the mirror reflects the light toward the diffraction grating, and the light breaks up due to collision with the grating. The decomposed light is irradiated on the detector by the second mirror and finally the Raman spectrum of the sample is obtained (Schmid and Dariz, 2019).

1481-Raman and 380-Raman are two common methods of Raman spectroscopy. 1481-Raman is based on the CH_2 bending vibrations in cellulose molecules and is intended for the analysis of pure cellulose, while 380-Raman can be used for both pure and noncellulosic materials (Foster et al., 2018). The degree of crystallinity can be calculated from the ratio of the peak intensity (I_{380}/I_{1096}) in the Raman spectrum for cellulosic materials (Agarwal et al., 2010). The estimation of crystallinity by 380-Raman uses a sample prepared by making a pellet with an approximate weight of 100 mg from nanocellulose powder with enough pressure to compress the sample to maintain its shape and appearance. In wet samples, a small amount of suspension with a concentration of 5 wt% is added to the sample tube. The samples are analyzed through FT-Raman and the spectrum is recorded with the signal-to-noise ratio. The signal-to-noise improvement is achieved by increasing the sampling time or the laser power, which depends on the state of the material (Foster et al., 2018; Agarwal et al., 2010). The crystallinity of cellulose is calculated using equation 6:

$$X_{380-Raman} = \frac{\frac{I_{380}}{I_{1096}} - 0.0286}{0.0065} \tag{6}$$

I_{380}/I_{1096} is obtained by calculating the height of the peaks at 380 and 1096 cm^{-1}.

6.2.6 Mechanical Properties

Bacterial cellulose has excellent mechanical properties due to the high strength and hardness of the cellulose molecular chain and is considered for application in various fields.

The choice of the best mechanical analysis method depends on the type of sample structure. Static tensile testing, dynamic mechanical testing, and viscoelastic measurement are among the mechanical analysis techniques that are used to investigate the mechanical properties of BC (Foster et al., 2018).

One of the important points in measuring the mechanical properties of composites made of BC is the use of standard analysis methods. ASTM D3039/D3039M-14, ASTM D 638–01, and ISO 527–1 are the standard methods for measuring the tensile properties of composites consisting of BC (Foster et al., 2018; Iwamoto et al., 2011). The tensile strength method is mainly used for composite films and fibers, but the viscoelastic properties of the sample cannot be evaluated. To investigate the viscoelastic properties of the sample, it is recommended to use dynamic mechanical testing, as discussed in the following section.

Dynamic mechanical testing is a useful method to check the viscoelastic mechanical properties of samples, including tensile strong modulus and tensile lose modulus. In this method, the modulus of composite samples can be changed by removing or adding a chemical regulator that can interfere with the hydrogen bonds of nanocellulose. In addition, the possibility of analyzing wet samples through dynamic mechanical testing was reported (Foster et al., 2018; Capadona et al., 2008).

Hydrogels have viscoelastic properties hence due to the hydrogel properties of bacterial cellulose, the measurement of the mechanical properties of hydrogels and cellulose-based hydrogels will be explained.

In hydrogel samples, viscoelastic analyses such as shear stress, modulus, hydration, single or cyclic compression tests, and tensile tests are time dependent. The samples are prepared in the form of discs and then placed in distilled water for 24 h to reach sufficient swelling. The compression test is another method of mechanical analysis for cellulose-based hydrogels. In this test, the hydrogel samples are placed between two impermeable plates and compressed (Foster et al., 2018).

6.2.7 Water Vapor Transmission Rate

Water vapor transmission rate (WVTR) is one of the most important parameters that are taken into consideration in investigating the properties of hydrogel wound dressings. This shows the ability to pass water vapor and the evaporation power of the hydrogel wound dressing. A circular wound dressing sample with a diameter of 9 cm is prepared and tested according to standard 9. The samples are placed on the special screen of the device for measuring the passage of steam through the fabric and sealed with glue. Then the weight of the sample is measured every hour and repeated for 5 to 6 h. Finally, the weight changes of the sample are calculated according to the drawing time and the water vapor passing rate is measured using equation 7:

$$WVTR = 24M / At \tag{7}$$

where M is the loss in mass of the sample over time, t is the time between equilibrium and end of the test (5–6 h) and A is the area of the sample (Jinka et al., 2013).

6.2.8 Water Absorption

The percentage of water absorption and swelling can describe the amount of water absorbed by the sample. To measure the amount of water absorption, the prepared samples with the same weight are immersed in distilled water for 24 h and after removing the extra water, the samples are weighed and considered as the weight of the wet sample. The samples are then dried and considered as dry weight. The water absorption of the samples is calculated through equation 8:

$$Water\,absorption\% = \left(W_w - W_d\right) / W_d \times 100 \tag{8}$$

where W_w is the wet weight and W_d is the dry weight of the sample (Feng et al., 2015).

6.2.9 Health/Safety Issues Characterizations

6.2.9.1 Cytotoxicity

In vitro cytotoxicity tests can show the toxicity ratio of a substance and the mechanism of action. There are different methods to perform cytotoxicity assay. MTS (3-(4,5-dimethylthiazol-2-yl)-5-(3-carboxymethoxyphenyl)-2-(4-sulfophenyl)-2H-tetrazolium), tetrazolium-based assays (MTT (3-(4,5-dimethylthiazol-2-yl)2,5-diphenyltetrazolium bromide), XTT (2,3-bis-(2-methoxy-4-nitro-5-sulfophenyl)-2H-tetrazolium-5carboxanilide)), lactate dehydrogenase assay, neutral red uptake, trypan blue, and alamar blue are the toxicity test methods. Among them, MTT is the most common test to determine BC toxicity. Therefore, the MTT method is described in detail.

First, 100 mL of the cell culture medium containing 90 mL DMEM (1X) + GlutMAXTM (GibcoTM Co) and 10 mL fetal calf serum is prepared. After trypsinization of normal fibroblast cells of the third passage produced from the foreskin of the neonate and cell count, 300 μL of culture medium containing 5000 cells is added to each well of the 96-well microplate and incubated for 48 h at 37°C with 5% carbon dioxide. In the next step, samples are cut into 1 × 1 inch squares and sterilized in an autoclave at 121°C and pressure 15 lb/inch^2 for 20 min. In addition, 2 mL of cell culture medium is added to the sterilized samples and incubated for 24 h at 37°C. Subsequently, the culture medium inside each well of the 96-well microplate is proliferated for 48 h and evacuated slowly by the sampler, and replaced with 200 μL of the adjoining food and the microplate then incubated at 37°C for 24 h with 5% carbon dioxide. The liquid on the cells is again replaced with fresh food and placed inside the incubator for 24 h. Then, 50 μL of 0.5 mg/mL MTT (3-[4,5-dimethylthiazol-2yl]-2,5-diphenyltetrazolium bromide) solution in PBS (phosphate-buffered saline) is added to each well and incubated for 4 h at 37°C with 5% carbon dioxide to form formazan crystals. Then the liquid on the cells is slowly evacuated and 200 μL of dimethylsulfoxide (DMSO) is added to each well. After 10 min of incubation at 37°C, the absorbance of each well is recorded by the microplate reader device at 570 nm and data are analyzed to measure the number of living cells. For each sample, three duplicate houses are considered and the average absorption of each sample is obtained after reducing the mean blank absorption. Three wells are only used to culture cells. The rate of cytotoxicity is obtained from equation 9:

$$Cell\,Viability\% = \frac{abs_{sample}}{abs_{control}} \times 100 \tag{9}$$

where abs_{sample} is the absorption rate for the sample and $abs_{control}$ is the control absorption rate (Fadaker Sarkandi et al., 2022; Ghayempour et al., 2016).

6.2.9.2 Wound Healing

Bacterial cellulose has hydrogel properties and can be applied in the field of medicine, especially as a wound dressing. Therefore, one of the analyses, in this field, is the wound healing assay described here in detail.

The evaluation of the wound healing of the sample is performed using a scratch scan on human primitive fibroblasts. In this test, proper extracellular matrix (ECM) for fibroblasts (10 mg/mL fibronectin as a control), is placed on the 60-mm plates and incubated at 4°C overnight. After removing the unbound ECM from the plates, the plates are filled with 3 mL of bovine serum albumin 2 mg/mL for 1 h at 37°C. Then the plates are washed with phosphate buffered saline (PBS), and filled with 3–5 mL media before placing the cell. In the following step, the cells are washed by PBS, the versene containing trypsin is added to the cells and mixed with serum medium. A portion of the cell suspension is taken to measure the number of cells. Cells are cultured in a 60-mm plate to form a cell layer. Using the tip of the micropipette, a scrape is created on the cell layer in a straight line and then washed using PBS. The scratched edge is washed with 1 ml of growth medium and replaced with 100 μl and 600 μL of the evaluated samples. The initial scratch image is recorded using a microscope. The plate is kept at 37°C for 24 h in an incubator. Then a second image was obtained by aligning the plate as shown in the first image. Using the images, the distance between the scratches can be measured. The migration rate of the cells into the scratch after 24 h is obtained from equation 10:

$$M_t\ \% = (d_0 - d_t)/d_0 \times 100 \qquad (10)$$

where d_0 is the distance between scratch at zero time and d_t is the distance between scratch at time t [65, 66].

6.3 STORAGE OF BACTERIAL CELLULOSE

Bacterial cellulose contains about 90% water due to its fibrous network structure and hydrogel property which indicates its porosity and ability to absorb and store water (Korhonen et al., 2011). The absorption and retention of water are unique properties of BC, and water must be removed for most applications. Bacterial cellulose is used in various fields, including in medicine as medical pads and wound and burn dressing, food and paper industries, diaphragm, and electroacoustic transducers (Barud et al., 2008; Zhang et al., 2011). Therefore, choosing the drying method for BC is very important because structure (porosity and morphology), mechanical strength, performance, and final applications are affected by the drying method (Illa et al., 2019).

Different methods are used to dry BC, including air drying (room temperature), oven drying, and freeze drying (Zhang et al., 2011). In short, freeze drying is a process in which the ice inside a frozen product turns into steam without any melting process. This takes place at a very low temperature and pressure called sublimation, resulting in a very high-quality product. The purpose of freeze drying is to produce a product with a long shelf life that can easily be mixed with water again without changing its texture. These factors are also dependent on the type of packaging and storage conditions. The advantage of freeze drying is less damage to heat-sensitive products than caused by drying is carried out at a low temperature (Barley, 2009; Chen and Wang, 2007).

In the air and oven drying methods, the microporous structures of the wet material are lost, while in the freeze drying method, the morphology and porosity of the fibers are preserved due to low surface tension (Feng et al., 2015; Tang et al., 2010; Al-Shamary and Al-Darwash, 2013). The drying method selected depends on the application of the sample. For instance, Classen et al. showed that the BC layer dried by freeze drying is useful for the application of wound dressings because the gas permeability is twice that of samples dried by air (Clasen et al., 2006). The effect of drying method on the morphology, porosity, crystallinity, chemical structure, and mechanical properties of the BC layer are discussed and their final properties compared.

6.3.1 EFFECT OF DRYING METHOD ON MORPHOLOGY

A study of the morphology of BC dried by air, oven, and freeze-drying shows that the samples have a typical shape and the diameter of the fibers is less than 100 nm using all three methods. Moreover,

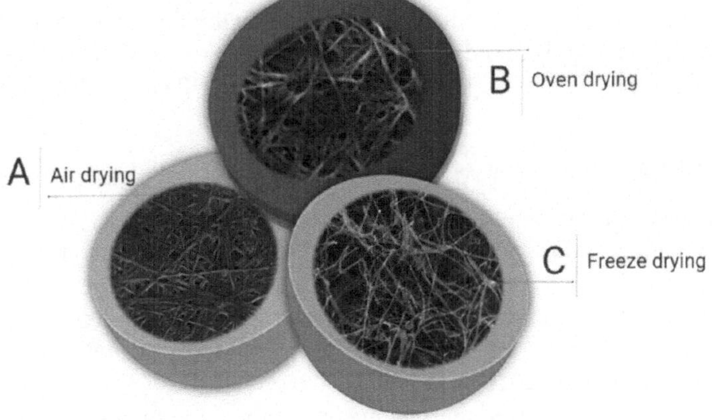

FIGURE 6.3 Field emission scanning electron microscope images of bacterial cellulose layers dried by (a) air drying, (b) oven drying, and (c) freeze drying.

the various drying methods affect the thickness of the layer; thus, bacterial nanocellulose dried by freeze-drying is the thickest and the sample dried in hot air is the thinnest (Zhang et al., 2011). Lack of sufficient entanglement resulting in more internal cavities in the structure of the layer dried by freeze-drying was reported, while the layers dried by oven and hot air have a more compact structure. It should be noted that the cavity in the samples dried by oven is more than in the samples dried in hot air. When the BC is dried by air, the moisture is lost slowly, resulting in the formation of a compact structure, while the drying is done faster by the oven leading to a defective structure. The moisture in the BC membrane turns into ice and then turns into gas under a vacuum in freeze drying, preserving the structure and appearance of the sample (Illa et al., 2019; Zhang et al., 2011). FESEM images of BC dried by air, oven, and freeze drying are shown in Figure 6.3.

The drying method is effective on the fiber diameter distribution. In general, the sample dried by freeze drying has a larger fiber diameter than the oven-dried sample. The increase in hydrogen bonding between the chains and also the increase in compressive force due to the evaporation of water cause a decrease in the diameter of the fibers dried in the oven method. The freezing of water during freeze drying reduces the movement of the chain and the hydrogen bonds, which ultimately results in much fewer sublimation forces.

Also, the specific surface area as well as the volume of the pores in samples dried by oven and freeze-drying are compared and investigated. In the oven-dried samples, the surface area decreases due to the collapse of the porous network structure and also the condensation of the network. The calculation of the pore size showed that in oven drying, the pore diameter is lower than 100 nm, while in freeze drying the pore size increased between 50 and 350 nm.

The presence of larger pores, macropores, in the freeze-dried sample is due to low surface tension. Meanwhile, the evaporation forces in oven drying cause the collapse of the pore structure and decrease the space between the fibers. Decreasing the fiber diameter in oven drying leads to a dense fibrous network, as a result of the lower surface area and the pore volume (Illa et al., 2019; Clasen et al., 2006).

6.3.2 Chemical Analysis by Fourier Transform Infrared Spectroscopy

Based on the study comparing drying methods, the spectra of all three samples are similar and the absorption peaks of cellulose are clearly defined. The peak of a sample obtained by freeze drying at 3345 cm^{-1} is more intense than the other two samples due to the OH stretching vibration. The higher stretching vibration is due to the more open structure and greater spacing of the cellulose chains, which ultimately leads to weak hydrogen bonds in freeze-dried layers. This is the reason for the more intense peak at 1055 cm^{-1}, related to C-OH stretching vibration, than in the samples dried by other methods. There is no significant difference between the spectra of the two methods of air drying and oven drying due to the similar structure of the samples (Illa et al., 2019; Zhang et al., 2011).

6.3.3 Degree of Crystallinity

The studies conducted on the x-ray diffraction pattern of BC layers dried by different methods show that the drying process does not affect the crystal structure due to maintaining the monoclinic structure of cellulose, although the diffraction intensity is different. The degree of crystallinity in the samples dried by freeze drying, air drying, and oven drying are 58.2%, 73.9%, and 63.8%, respectively, which shows that the BC dried by hot air has a higher degree of crystallinity than the other two samples (Zhang et al., 2011).

In the drying by hot air and oven, thermal energy leads to the organization of the amorphous region, and as a result crystallinity increases. In addition, the crystal size in oven-dried samples is smaller than in the freeze-dried samples (Illa et al., 2019; Zhang et al., 2011).

6.3.4 Mechanical Properties

The physical and mechanical properties of BC are also affected by different drying methods. The Young's modulus of BC layers that were treated with alkali and then subjected to heat and pressure increased dramatically (Chen et al., 2021; Markiewicz et al., 2004). Therefore, in this section, a comparison of the mechanical properties of BC films dried by three different methods is discussed.

Among the methods for drying BC, the hot air method shows the highest strength. High crystallinity, homogeneous and compact structure, and strong intermolecular bonds of the BC increased the strength and Young's modulus. On the other hand, the lowest strength is exhibited in BC dried by freeze drying. As previously discussed, the samples obtained from freeze drying have low crystallinity and weak intermolecular hydrogen bonds due to the cavity and layered structure as well as the presence of large spaces. As a result, the strength and Young's modulus are the lowest compared to the hot air and oven drying methods. The presence of voids and defective structure in the oven-dried layer and lower hydrogen bonds and crystallinity leads to a decrease in strength (Illa et al., 2019; Zhang et al., 2011; Sehaqui et al., 2011; South et al., 2005).

6.4 CONCLUSIONS

In recent years, research and development in the field of cellulosic nanomaterials, especially microbial cellulose, has received much attention and grown significantly in response to the limited availability of natural resources. For this reason, it is necessary to describe and analyze bacterial cellulose without any confusion and problems. In this regard, an attempt was made to explain the necessary chemical, mechanical, morphological, and other physical properties of bacterial cellulose to increase the accuracy of analysis and measurement. In addition, all the drying methods for bacterial cellulose are fully described. Because of the wide use of bacterial cellulose in various fields and given that the properties of bacterial cellulose are influenced by the drying method, selection of the appropriate method is required depending on the final application. Finally, the chapter summarized and highlighted the important and practical work of experts and researchers to facilitate academic activities.

6.5 REFERENCES

Afanasyev, S., T. Kychkina, and L. Savvinova. *Scanning electron microscope (advantages and disadvantages).* Colloquium-Journal, 2019. **2–2**(26): p. 25–27. Голопристанський міськрайонний центр зайнятості.

Agarwal, U.P., R.S. Reiner, and S.A. Ralph, *Cellulose I crystallinity determination using FT: Raman spectroscopy: Univariate and multivariate methods.* Cellulose, 2010. **17**(4): p. 721–733.

Akhtar, K., S. Khan, S. Khan, and A.M. Asiri, *Scanning electron microscopy: Principle and applications in nanomaterials characterization*, in *Handbook of Materials Characterization.* 2018: Springer. p. 113–145.

Al-Shamary, E.E., and A.K. Al-Darwash, *Influence of fermentation condition and alkali treatment on the porosity and thickness of bacterial cellulose membranes.* Tojsat, 2013. **3**(2): p. 194–203.

Angela, C., *Effect of PH, molasses, and caffeine concentration on the yield and the properties of bacterial celulose biofilms.* 2019, Indonesia International Institute for Life Sciences.

Ayache, J., L. Beaunier, J. Boumendil, G. Ehret, D. Laub, *Sample Preparation Handbook for Transmission Electron Microscopy: Techniques.* 2010. **2**: Springer Science & Business Media.

Barley, J., *Basic principles of freeze drying.* SP Scientific. Available online at: www.spscientific.com/freeze-drying-lyophilization-basics, 2009.

Barud, H., R. Assunção, M. Martines, J. Dexpert-Ghys, R.F.C. Marques, Y. Messaddeq, and S.J.L. Ribeiro, *Bacterial cellulose: Silica organic: Inorganic hybrids.* Journal of Sol-Gel Science and Technology, 2008. **46**(3): p. 363–367.

Bellitto, V., *Atomic Force Microscopy: Imaging, Measuring and Manipulating Surfaces at the Atomic Scale.* 2012: BoD-Books on Demand.

Briggs, D., *Handbook of X-Ray and Ultraviolet Photoelectron Spectroscopy.* 1977: Heyden.

Buckley, A., J.I. Goldstein, D.E. Newbury, P. Echlin, D.C. Joy, A.D. Romig, Jr., C.E. Lyman, C. Fiori, and E. Lifshin 1992. *Scanning electron microscopy and X-Ray microanalysis: A text for biologists, materials scientists, and geologists, xviii+ 820 pp. New York, London: Plenum Press. Price US $49.50 (hard covers). ISBN 0 306 44175 6.* Geological Magazine, 1993. **130**(3): p. 402–403.

Capadona, J.R., K. Shanmuganathan, D.J. Tyler, S.J. Rowan, and C. Weder, *Stimuli-responsive polymer nanocomposites inspired by the sea cucumber dermis.* Science, 2008. **319**(5868): p. 1370–1374.

Carbon, H., *Nitrogen determination using the PerkinElmer 2400 series II CHNS.* 2011, O Analyzer, PerkinElmer.

Carter, C.B., and D.B. Williams, *Transmission Electron Microscopy: Diffraction, Imaging, and Spectrometry.* 2016: Springer.

Chatterjee, S., S.S. Gadad, and T.K. Kundu, *Atomic force microscopy.* Resonance, 2010. **15**(7): p. 622–642.

Chen, G., and W. Wang, *Role of freeze drying in nanotechnology.* Drying Technology, 2007. **25**(1): p. 29–35.

Chen, S.-Q., et al., *The influence of alkaline treatment on the mechanical and structural properties of bacterial cellulose.* Carbohydrate Polymers, 2021. **271**: p. 118431.

Chen, Y., C. Zou, M. Mastalerz, S. Hu, C. Gasaway, and X. Tao, *Applications of micro-fourier transform infrared spectroscopy (FTIR) in the geological sciences: A review.* International Journal of Molecular Sciences, 2015. **16**(12): p. 30223–30250.

Cirtiu, C.M., A.F. Dunlop-Briere, and A. Moores, *Cellulose nanocrystallites as an efficient support for nanoparticles of palladium: Application for catalytic hydrogenation and heck coupling under mild conditions.* Green Chemistry, 2011. **13**(2): p. 288–291.

Clasen, C., B. Sultanova, T. Wilhelms, P. Heisig, W. Kulicke, and C. Clasen, *Effects of different drying processes on the material properties of bacterial cellulose membranes*, in *Macromolecular Symposia.* 2006: Wiley Online Library.

Davidson, M.W., and M. Abramowitz, *Optical microscopy*, in *Encyclopedia of Imaging Science and Technology*, 2002. **2**(1106–1141). p. 120.

Dey, A., S. Mandal, S. Bhandari, C. Pal, J. Orasugh, and D. Chattopadhyay, *Characterization methods*, in *Fiber-Reinforced Nanocomposites: Fundamentals and Applications.* 2020: Elsevier. p. 7–67.

Eaton, P., and P. West, *Atomic Force Microscopy.* 2010: Oxford university press.

Elipe, M.V.S., *Advantages and disadvantages of nuclear magnetic resonance spectroscopy as a hyphenated technique.* Analytica Chimica Acta, 2003. **497**(1–2): p. 1–25.

Elmer, P., *FT-IR spectroscopy attenuated total reflectance (ATR).* Technical Note, 2005. **27**(11).

Fadakar Sarkandi, A., M. Montazer, T. Harifi, and M.R. Mahnaz, *Innovative preparation of bacterial cellulose/silver nanocomposite hydrogels: In situ green synthesis, characterization, and antibacterial properties.* Journal of Applied Polymer Science, 2021. **138**(6): p. 49824.

Fadakar Sarkandi, A., M. Montazer, and M. Mahmoudi Rad, *Oxygenated-bacterial-cellulose nanofibers with hydrogel, antimicrobial, and controlled oxygen release properties for rapid wound healing.* Journal of Applied Polymer Science, 2022. **139**(16): p. 51974.

Feng, X., N. Ullah, X. Wang, X. Sun, C. Li, Y. Bai, L. Chen, and Z. Li, *Characterization of bacterial cellulose by Gluconacetobacter hansenii CGMCC 3917.* Journal of Food Science, 2015. **80**(10): p. E2217–E2227.

Foster, E.J., Robert J. Moon, Umesh P. Agarwal, Michael J. Bortner, Julien Bras, Sandra Camarero-Espinosa, Kathleen J. Chan, Martin J.D. Clift, Emily D. Cranston, Stephen J. Eichhorn, Douglas M. Fox, Wadood Y. Hamad, Laurent Heux, Bruno Jean, Matthew Korey, World Nieh, Kimberly J. Ong, Michael S. Reid, Scott Renneckar, Rose Roberts, Jo Anne Shatkin, John Simonsen, Kelly Stinson-Bagby, Nandula Wanasekara, and Jeff Youngblood, *Current characterization methods for cellulose nanomaterials.* Chemical Society Reviews, 2018. **47**(8): p. 2609–2679.

Gao, W.-H., K.-F. Chen, R.-D. Yang, F. Yang, and W.-J. Han, *Properties of bacterial cellulose and its influence on the physical properties of paper.* BioResources, 2011. **6**(1): p. 144–153.

Ghayempour, S., M. Montazer, and M.M. Rad, *Simultaneous encapsulation and stabilization of aloe vera extract on cotton fabric for wound dressing application.* RSC Advances, 2016. **6**(113): p. 111895–111902.

Griffiths, J., *Secondary ion mass spectrometry.* Analytical chemistry (Washington, DC), 2008. **80**(19): p. 7194–7197.

Gritsch, M., C. Brunner, K. Piplits, H. Hutter, P. Wilhartitz, A. Schintlmeister, H.P. Martinz, and M. Gritsch, *Application of scanning SIMS techniques for the evaluation of the oxidation behavior of high-purity molybdenum.* Fresenius' Journal of Analytical Chemistry, 1999. **365**(1): p. 188–194.

Haugstad, G., *Atomic Force Microscopy: Understanding Basic Modes and Advanced Applications.* 2012: John Wiley & Sons.

Hill, S.J., *Inductively Coupled Plasma Spectrometry and its Applications.* 2008: John Wiley & Sons.

Huang, M., and G.M. Hieftje, *Simultaneous measurement of spatially resolved electron temperatures, electron number densities and gas temperatures by laser light scattering from the ICP.* Spectrochimica Acta Part B: Atomic Spectroscopy, 1989. **44**(8): p. 739–749.

Illa, M.P., C.S. Sharma, and M. Khandelwal, *Tuning the physiochemical properties of bacterial cellulose: Effect of drying conditions.* Journal of Materials Science, 2019. **54**(18): p. 12024–12035.

Iwamoto, S., A. Isogai, and T. Iwata, *Structure and mechanical properties of wet-spun fibers made from natural cellulose nanofibers.* Biomacromolecules, 2011. **12**(3): p. 831–836.

Jinka, S., R. Behrens, C. Korzeniewski, V. Singh, A. Arunachalam, Siva Parameswaran, Gopal Coimbatore, R.J. Kendall, R. Wolf, and S. Ramkumar, *Atmospheric pressure plasma treatment and breathability of polypropylene nonwoven fabric.* Journal of Industrial Textiles, 2013. **42**(4): p. 501–514.

Johansson, L.S., and J. Campbell, *Reproducible XPS on biopolymers: Cellulose studies.* Surface and Interface Analysis: An International Journal Devoted to the Development and Application of Techniques for the Analysis of Surfaces, Interfaces and Thin Films, 2004. **36**(8): p. 1018–1022.

Kaushik, Madhu., Carole Fraschini, Grégory Chauve, Jean-Luc Putaux, and Audrey Moores, *Transmission electron microscopy for the characterization of cellulose nanocrystals*, in *The Transmission Electron Microscope-Theory and Applications.* 2015: InTech. p. 130–163. doi:10.5772/60985.

Khan, H., A.S. Yerramilli, A. D'Oliveira, T.L. Alford, D.C. Boffito, and G.S. Patience, *Experimental methods in chemical engineering: X-ray diffraction spectroscopy—XRD.* The Canadian Journal of Chemical Engineering, 2020. **98**(6): p. 1255–1266.

Korhonen, J.T., M. Kettunen, R.H. Ras, and O. Ikkala, *Hydrophobic nanocellulose aerogels as floating, sustainable, reusable, and recyclable oil absorbents.* ACS Applied Materials & Interfaces, 2011. **3**(6): p. 1813–1816.

Krystynowicz, A., W. Czaja, A. Wiktorowska-Jezierska, M. Gonçalves-Miśkiewicz, M. Turkiewicz, and S. Bielecki, *Factors affecting the yield and properties of bacterial cellulose.* Journal of Industrial Microbiology and Biotechnology, 2002. **29**(4): p. 189–195.

Kumar, A., Y. Negi, V. Choudhary, and N. Bhardwaj, *Characterization of cellulose nanocrystals produced by acid-hydrolysis from sugarcane bagasse as agro-waste.* Journal of Materials Physics and Chemistry, 2014. **2**(1): p. 1–8.

Kurukavak, C.K., and T. Yilmaz, *Characterization of bionanocomposites*, in *Bionanocomposites in Tissue Engineering and Regenerative Medicine.* 2021: Elsevier. p. 45–90.

Lis, G.P., M. Mastalerz, A. Schimmelmann, M. Lewan, A. Stankiewicz, and G. Lis, *FTIR absorption indices for thermal maturity in comparison with vitrinite reflectance R0 in type-II kerogens from Devonian black shales.* Organic Geochemistry, 2005. **36**(11): p. 1533–1552.

Markiewicz, E., B. Hilczer, and C. Pawlaczyk, *Dielectric and acoustic response of biocellulose.* Ferroelectrics, 2004. **304**(1): p. 39–42.

Mayeen, A., L. Shaji, A. Nair, and N. Kalarikkal, *Morphological characterization of nanomaterials*, in *Characterization of Nanomaterials*. 2018: Elsevier. p. 335–364.

McHenry, M., and D. Laughlin, *Characterization of Materials*. 2012: Wiley.

Michen, B., C. Geers, D. Vanhecke, C. Endes, B. Rothen-Rutishauser, S. Balog, and A. Petri-Fink, *Avoiding drying-artifacts in transmission electron microscopy: Characterizing the size and colloidal state of nanoparticles*. Scientific Reports, 2015. **5**(1): p. 1–7.

Mohammadkazemi, F., K. Doosthoseini, E. Ganjian, and M. Azin, *Manufacturing of bacterial nano-cellulose reinforced fiber: Cement composites*. Construction and Building Materials, 2015. **101**: p. 958–964.

Nasrollahzadeh, M., M. Atarod, M. Sajjadi, S.M. Sajadi, and Z. Issaabadi, *Plant-mediated green synthesis of nanostructures: Mechanisms, characterization, and applications*, in *Interface Science and Technology*. 2019: Elsevier. p. 199–322.

Neto Flauzino, W.P., Jean-Luc Putaux, Marcos Mariano, Yu Ogawa, Harumi Otaguro, Daniel Pasquini, and Alain Dufresne, *Comprehensive morphological and structural investigation of cellulose I and II nanocrystals prepared by sulphuric acid hydrolysis*. RSC Advances, 2016. **6**(79): p. 76017–76027.

Özdöl, V.B., V. Srot, and P.A. van Aken, *Sample preparation techniques for transmission electron microscopy*, in *Handbook of Nanoscopy*. 2012: Wiley-VCH Verlag GmbH & Co. KGaA. p. 473–498.

Pang, J., et al., *Hydrolysis of cellulose into glucose over carbons sulfonated at elevated temperatures*. Chemical Communications, 2010. **46**(37): p. 6935–6937.

Paramasivan, M., T.S.S. Kumar, and T. Chandra, *Microbial synthesis of hydroxyapatite-nanocellulose nanocomposites from symbiotic culture of bacteria and yeast pellicle of fermented kombucha tea*. Sustainability, 2022. **14**(13): p. 8144.

Park, S., J.O. Baker, M.E. Himmel, P.A. Parilla, and D.K. Johnson, *Cellulose crystallinity index: Measurement techniques and their impact on interpreting cellulase performance*. Biotechnology for Biofuels, 2010. **3**(1): p. 1–10.

Parker, G., *Encyclopedia of Materials: Science and Technology*, in K.H. Jürgen Buschow, Robert W. Cahn, Patrick Veyssière (Eds.) *Guide-Wave Optical Communications: Materials*. 2001: Elsevier. p. 3703–3707.

Reid, M.S., M. Villalobos, and E.D. Cranston, *Benchmarking cellulose nanocrystals: From the laboratory to industrial production*. Langmuir, 2017. **33**(7): p. 1583–1598.

Revin, V., E. Liyaskina, M. Nazarkina, A. Bogatyreva, and M. Shchankin, *Cost-effective production of bacterial cellulose using acidic food industry by-products*. Brazilian Journal of Microbiology, 2018. **49**: p. 151–159.

Schenzel, K., S. Fischer, and E. Brendler, *New method for determining the degree of cellulose I crystallinity by means of FT Raman spectroscopy*. Cellulose, 2005. **12**(3): p. 223–231.

Schmid, T., and P. Dariz, *Raman microspectroscopic imaging of binder remnants in historical mortars reveals processing conditions*. Heritage, 2019. **2**(2): p. 1662–1683.

Sehaqui, H., Q. Zhou, O. Ikkala, and L.A. Berglund, *Strong and tough cellulose nanopaper with high specific surface area and porosity*. Biomacromolecules, 2011. **12**(10): p. 3638–3644.

Shojaeiarani, J., D.S. Bajwa, and S. Chanda, *Cellulose nanocrystal based composites: A review*. Composites Part C: Open Access, 2021. **5**: p. 100164.

Siegbahn, K., *Electron spectroscopy for atoms, molecules, and condensed matter*. Reviews of Modern Physics, 1982. **54**(3): p. 709.

Singh, M.K., and A. Singh, *Characterization of Polymers and Fibres*. 2021: Elsevier.

Smith, E., and G. Dent, *Modern Raman Spectroscopy: A Practical Approach*. 2019: John Wiley & Sons.

Sodhi, R.N., *Time-of-flight secondary ion mass spectrometry (TOF-SIMS): Versatility in chemical and imaging surface analysis*. Analyst, 2004. **129**(6): p. 483–487.

South, M., S. Ozonoff, and W.M. McMahon, *Repetitive behavior profiles in asperger syndrome and high-functioning autism*. Journal of Autism and Developmental Disorders, 2005. **35**(2): p. 145–158.

Steyermark, A., *Quantitative Organic Microanalysis*. 2012: Elsevier.

Takegawa, A., M. Murakami, Y. Kaneko, and J. Kadokawa, *A facile preparation of composites composed of cellulose and polymeric ionic liquids by in situ polymerization of ionic liquids having acrylate groups*. Polymer Composites, 2009. **30**(12): p. 1837–1841.

Tang, W., S. Jia, Y. Jia, and H. Yang, *The influence of fermentation conditions and post-treatment methods on porosity of bacterial cellulose membrane*. World Journal of Microbiology and Biotechnology, 2010. **26**(1): p. 125–131.

Tonoli, G., et al., *Cellulose micro/nanofibres from eucalyptus kraft pulp: Preparation and properties*. Carbohydrate Polymers, 2012. **89**(1): p. 80–88.

Zhang, C.J., L. Wang, J. Zhao, and P. Zhu, *Effect of drying methods on structure and mechanical properties of bacterial cellulose films*, in *Advanced Materials Research*. 2011: Trans Tech Publ.
Zhang, Z., X. Meng, W. Cui, S.S. Ahmad, M.A. Kamal, and J. Zhang, *NMR: From molecular mechanism to its application in medical care*. Medicinal Chemistry, 2020. **16**(8): p. 1089–1098.
Zhou, W., and Z.L. Wang, *Scanning Microscopy for Nanotechnology: Techniques and Applications*. 2007: Springer Science & Business Media.
Zuo, J.M., and J.C. Spence, *Advanced Transmission Electron Microscopy*. 2017: Springer.

7 Chemical Modification of Bacterial Cellulose for Industrial Applications

Anupreet Kaur Sobti and Vasudha Bansal
Government Home Science College
(Affiliated with Panjab University, Chandigarh)
Chandigarh, Punjab and Haryana, India

Nitya Sharma
Indian Institute of Technology Delhi
New Delhi, India

7.1 INTRODUCTION

The most abundant, affordable, and easily accessible carbohydrate polymer in the world is cellulose, which is often obtained from plants or their byproducts (Sun, 2008). Cellulose is a linear homopolysaccharide with β-D-glucopyranose units joined together by β-1,4 glycosidic linkages (Figure 7.1). Because of its significant contribution to productivity, cellulose is a top biopolymer. To produce the pure material, this polymer must go through an unpleasant chemical process with harsh alkali and acid treatment, which generally branches with hemicellulose and lignin. Growing demand for plant cellulose derivatives increased the use of wood as a raw material, leading to deforestation and a global environmental problem (Park, Park, & Jung, 2003).

Cellulose is a class of carbohydrates with a high concentration of hydroxyl groups that persist as polymer chains (Ross, Mayer, & Benziman, 1991) and are widely used in the pulp and paper, pharmaceutical, and renewable fuel industries. The challenge in procuring pure cellulose as a substrate, however, is due to the plant cellulose's frequent association with lignin, pectin, hemicellulose, and other biogenic compounds. It has been discovered that, in addition to being the most significant structural component of the primary cell wall of green plants, cellulose exists in bacteria. Bacterial cellulose (BC) offers the highest level of purity, which has piqued the interest of numerous researchers in academia and in industrial sectors. BC typically comprises bundles of 7–8 nm-wide elementary nanofibrils formed at random to form fibrils with a width of about 100 nm and a ribbon-like structure. As a result, it offers a unique mix of qualities, encompassing biomimetic three-dimensional (3D) network, high water holding capacity, hydrophilicity, crystallinity, moldability into various shapes, and increased purity with lack of lignin and hemicellulose. These characteristics make this type of cellulose enticing for use in a variety of medical applications, including the development of artificial skin (especially for the recovery of burned skin), artificial blood vessels, coverings for nerve surgery, dura mater prostheses, topical dressings for severe wounds, arterial stent coatings, wound dressings, hemostatic material, electronic platforms, implants for cartilage and bone repair, and much more.

A. J. Brown initially identified BC in 1886 as the extracellular gelatinous fiber made by *Acetobacter xylinum*. Initially, BC was used extensively in the production of coconut gel, commonly known as nata (also called nata de coco). The process of fermenting coconut water by bacteria yielded this transparent jelly-like delicacy, which was first made famous in the Philippines in 1973. In several Asian nations, including Vietnam, Indonesia, and Japan, BC became increasingly popular; 90% of

DOI: 10.1201/9781003355434-7

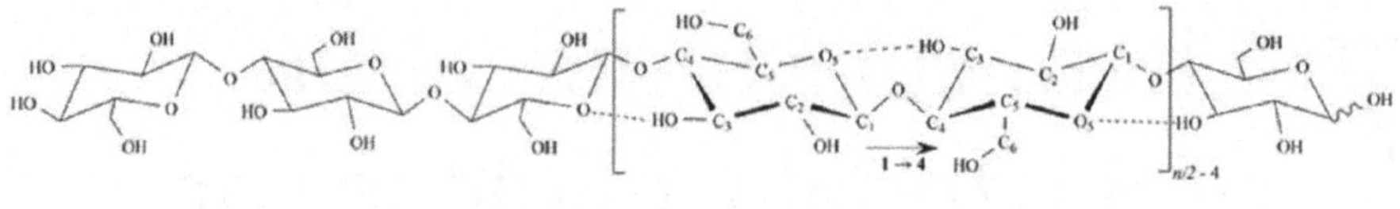

FIGURE 7.1 Chemical structure of the cellulose molecule.

the nata produced in the Philippines in 1993 was shipped to Japan, according to market historical records. Nata from British Columbia was made famous as a food with numerous health advantages, and this claim was important to its commercial success. Additionally, the simplicity of BC (as nata) production fueled the expansion of several small companies in rural southeast Asia. As a paper binding agent, a basis for clothes, and a substance for wound treatment, BC has started to permeate modern commercial and medical applications. Market interest in BC is anticipated to increase due to its widespread use.

Microorganisms synthesize BC, a structural carbohydrate. BC has the same chemical structure as plant cellulose, $(C_6H_{10}O_5)_n$, but BC contains nanoscale polymer fibers. BC is a metabolic byproduct produced extracellularly by bacteria. *Gluconacetobacter, Sarcina, Agrobacterium, Rhizobium, Rhodobacter*, and *Agrobacterium* are a few typical cellulose-producing bacterial strains that have been found (Moniri, Boroumand, & Azizi et al., 2017; Yin, Santos, & Auer et al., 2014; Aydin & Aksoy, 2014). The *Gluconacetobacter xylinus* species, formerly known as *Acetobacter xylinum* and now known as *Komagataeibacter xylinus*, is renowned for its capacity to generate BC on a large scale. BC from this strain has lately been employed with more sophisticated materials in biomedical and tissue engineering, as well as in many other industries, including food and packaging goods.

7.2 PRODUCTION OF BACTERIAL CELLULOSE

7.2.1 Fermentation Techniques

Only the *Gluconacetobacter* genus with the species name *hansenii* has been identified as a Gram-positive genus of acetic acid bacteria (AAB), and almost all the AAB species are Gram-negative (Hommel, 2014). Because it can produce BC on a commercial scale, *Gluconacetobacter xylinus,* one of the major AAB species, is frequently employed as a model organism to research BC biosynthesis (Morgan, Strumillo, & Zimmer, 2013). The ultrafine 3D nanofibril structure of BC fermented by AAB possesses exceptional crystallinity and purity.

7.2.2 Static Fermentation

For extracellular products, static (or surface) culture is a popular kind of fermentation. Due to the ease of use compared to other production methods, static culture has been widely used. This method uses shallow bottles or trays carrying the liquid growth media to cultivate bacteria, as shown in Figure 7.2 (a). After inoculation and fermentation at a temperature of 25°C to 30°C Celsius and a pH of 3 to 7, BC can be grown for a number of days. The surface of the medium eventually becomes coated with a floating layer of a gelatinous BC pellicle during a cultivation period of 5 to 20 days. The bacteria themselves will eventually be entrapped in the BC pellicle.

7.2.3 Submerged Fermentation

Static culture has some drawbacks despite being commonly used in BC production, primarily due to its lengthy cultivation period and low productivity. Additionally, during cultivation, bacteria are

FIGURE 7.2 Bacterial cellulose (BC) produced via static and agitated fermentation.

(a, b) BC pellicle formed at the air-liquid interface of the medium in a static fermentation process (a) and the purified BC pellicle with uniform texture (b). (c, d) BC pellets fully filled in the medium in an agitated fermentation process (c) and the purified BC pellets with irregular shapes. Reproduced from (Zhong, 2020). With permission.

subjected to unfavorable conditions (in terms of nutrients, oxygen, and population dispersion) during their growth cycle, which results in an uneven production of BC layer thickness. These problems cause a buried culture to emerge. Numerous studies have demonstrated the benefits of submerged fermentation over static culture, the main one being enhanced production. Submerged fermentation, however, also faces a number of challenges, including the emergence of cellulose-nonproducing strains (Matsutani, Ito, Azuma, et al., 2015), the generation of BC granules with irregular shapes, and the alteration of BC's physical properties (Singhsa, Narain, & Manuspiya, 2018). Additionally, the bacteria favor the synthesis of gluconic acid over the creation of cellulose under conditions of high rotational speed and hydrostatic pressures that result in the buildup of metabolites for self-defense (Liu, Zhong, Wu, et al., 2015). Numerous submerged fermentation techniques, such as stirred tank, rotating disc, and airlift bioreactors, have been used to produce BC (Table 7.1 and Table 7.2).

The benefits of submerged fermentation have been further enhanced by the use of a plastic composite support (PCS). PCS, a polypropylene and nutrient combination compost, promotes biofilm adherence and increases BC production (Cheng, Catchmark, & Demirci, 2009). With the success of the PCS biofilm reactor, other research based on this PCS biofilm reactor that utilized a PCS rotating disc bioreactor have been completed (PCS-RDB).

With a set of net plates arranged vertically in the reactor and an air distributor at the bottom, Wu et al. (Wu & Li, 2015) created a new airlift bioreactor that could produce a membrane-type BC. This bioreactor design generated an agitated BC film that resembled a membrane.

TABLE 7.1
Methods and Characteristics of Bacterial Cellulose (BC) Production (Gorgieva, 2019)

Method of BC Production	Characteristics of Cellulose and the Process
Static production	Most commonly used method at the lab scale. Duration of the process is up to two weeks. Cellulose is in the form of hydrogel sheet.
Production in shaking culture	Increased delivery of oxygen to bacteria. Might result in reduced genetic stability of bacteria and lower BC production. Production of cellulose of different particle sizes and various shapes (mainly of spherical structure). Suitable for economic-scale production.
Production in airlift bioreactor	Efficient oxygen supply with low power supply. Cellulose produced in pellet.
Production in rotating disc bioreactors	Production of homogeneous cellulose. Cellulose yield is compared to the static process.
Production in trickling bed reactor	Provides high oxygen concentration and low shear force. Produces BC in form of irregular sheets.

TABLE 7.2
Comparing Different Types of Bioreactors Employed for Bacterial Cellulose Production (Blanco Parte, Santoso, Chou et al., 2020)

Reactor Type	Advantage	Disadvantage
Stirred tank reactor	High cell concentration	High power demand, high shear stress
Airlift reactor	High cell concentration; relatively low shear stress	High energy requirement
Rotating disk reactor	High cell concentration; ease of ingredients incorporation	Semicontinuous production
Trickling bed reactor	Higher oxygen supply; better physical properties	Semicontinuous production; recovery issue

7.2.4 Chemical Modification in Bacterial Cellulose Fermentation

7.2.4.1 Culture Media

Economically considered, if culture material is produced via much less expensive sources, low productivity may be acceptable. For instance, it has been demonstrated that citrus peel and pomace waste from the beverage industry have potential as a BC culture medium. Additionally, the employed waste medium produced a BC yield of 5.7 g/L, which was higher than the typical BC yield in Hestrin-Schramm (HS) medium of only 3.9 g/L (Fan, Gao, He, et al., 2016). Waste beer yeast (WBY) has demonstrated that it can grow *Gluconacetobacter hansenii* CGMCC 3917 and produce BC on its own. In comparison to WBY that had not been ultrasonically treated, the yield of BC was increased by six times, and more significantly, by two times when BC was produced using standard chemical media (Lin, Lopez-Sanchez, Li, et al., 2014).

These investigations show that alternative media with a complex natural composition are just as efficient and economical as the conventional synthetic medium defined by chemicals. It is not surprising that cellulosic materials are being investigated for their potential as a feedstock for BC synthesis as fermentable sugars are used as the carbon source in BC production. For instance, after receiving an enzymatic treatment, fiber sludge from the sulfate (SAFS) and sulfite (SIFS) processes was utilized as a feedstock to create BC, yielding 11 g/L and 10 g/L, respectively (Çakar,

Ozer, Aytekin, et al., 2014). The monetary value of using low-value trash to create high-value commodities is unquestionably far larger. Algal biomass is another possible source of carbon. In one of the investigations, glucose was produced for BC by hydrolyzing starch from the algae *Chlorella vulgaris*. The algae were initially depleted of nitrogen to raise its starch content. To create BC, the hydrolysate was then added to the fermentation medium under static conditions. In comparison to 1.202 g/L of BC produced by the same procedure using ordinary glucose as a carbon source, our process produced 1.104 g/L of BC (Cavka, Guo, Tang, et al., 2013). There has also been much research on fruits and related byproducts. The possibility of fruit peels, rotten fruit (Uzyol & Sacan, 2017), and waste from the manufacturing of fruit juice as BC feedstock has been investigated. Due to the significant amount of fruit that is harmed during shipping and the sporadic bumper harvests, fruits are occasionally allowed to perish in the field. If this method can turn fruit attrition into high-value BC, farmers and BC producers could both gain from it. Numerous byproducts of the wine business have the potential to be produced in BC. Thin stillage added to a conventional HS medium has been shown to generate 2.5 times as much BC as an HS medium without supplementation (Jozala, Pertile, dos Santos, et al., 2015). Because the waste from candied jujube is so full of organic nutrients, it may seriously harm the environment if it is dumped out into the environment. However, this effluent can produce 2.25 g/L of BC when heated and acidified before being diluted into varied proportions. Also, lipid fermentation wastewater can create 0.659 g/L of BC by day 5 and a decreased 30% of Chemical Oxygen Demand (COD) from the wastewater when infected with *G. xylinus* CH001 (Wu & Liu, 2012; Li, Wang, Hua, et al., 2015; Huang, Guo, Xiong, et al., 2016).

7.2.4.2 Carbon and Nitrogen Sources

Different carbon sources are required by various cellulose-producing bacterial strains. For example, bacteria in the *Komagataeibacter* genus have a wide range of optimal carbon sources, including glucose, sucrose, and fructose (Mohite, Salunke, & Patil, 2013; Molina-Ramırez, Castro, Osorio, et al., 2017). Glycerol and sucrose were the two carbon sources that *Komagataeibacter sucrofermentans* DSM 15973 responded to the best (Tsouko, Kourmentza, Ladakis, et al., 2015). Similar research has been done on nitrogen sources in an effort to find a low-cost, high-productivity nitrogen source. For instance, the use of decaying fruits as a carbon source and milk whey as a source of nitrogen produced more BC than the reference HS medium (Jozala, Pertile, dos Santos, et al., 2015).

7.2.4.3 Magnitude of pH

Numerous studies have been done on the significance of pH during the synthesis of BC. A pH of less than 7 is ideal for the *Komagataeibacter* genus. However, the pH value gradually drops during fermentation and frequently goes below the ideal pH for BC production. This is due to a variety of metabolic pathways that are currently in use, including the oxidation of ethanol to acetic acid and the conversion of glucose to gluconic acid (Lin, Loira, Catchmark, et al., 2013). Colony forming units (CFUs) on HS-agar plates were used to study the influence of pH on the *Gluconacetobacter medellinensis* strain ID13488, and the results show that acidic environments exhibit superior cell viability than neutral pH (Urbina, Hernandez-Arriaga, Eceiza, et al., 2017). Although the majority of strains generally produce more BC in surroundings that are a little bit acidic, at least one strain (*K. intermedius* FST213-1) demonstrated superior BC production in basic conditions. The highest BC production was at pH 8, and up to pH 9, when the first study reported BC production, cellulose-producing ability had not been established (Lin, Huang, Hsu, et al., 2016)[32]. Although the majority of strains prefer an acidic environment, the buildup of gluconic acid appears to have a negative impact on the bacteria's productivity. Better production rates and the yield of BC during conversion are obtained when an acetate buffer is added to the medium. The amount of BC produced in the 200 mM, pH 4.75, acetate-buffered medium was 3.56 g/L, which was larger than the amounts produced in the Yeast Extract Peptone Dextrose (YPD) (0.66 g/L) and HS (1.23 g/L) media (Kuo, Chen, Liou, et al., 2016).

7.2.4.4 Oxygen Level

It is known that the *Komagataeibacter* genus, which produces BC, is an exclusively aerobic microbe (Saichana, Matsushita, Adachi, et al., 2015). This implies that they must have access to oxygen in order to thrive and produce BC. In the bacterial cells, adenosine triphosphate (ATP) is activated by oxygen as a precursor. Surface area, for example, has a direct relationship with the amount of dissolved oxygen (DO) present in the growth medium. In an airlift bioreactor, DO was assumed to be the limiting element, and it was hypothesized that cellulose synthesis is more closely tied to oxygen uptake than to medium nutrients. In a different experiment using a variety of media, including the HS medium and an apple residue/sugarcane medium with a static culture, depletion of DO was not seen during the 13-day cultivation period, indicating that DO may not have been the limiting factor in this instance. A pBla-VHb-122 plasmid was used to introduce the Vitreoscilla haemoglobin (VHb)-producing gene *vgb* into *G. xylinus* to better examine how oxygen tension affects cell proliferation and BC synthesis. VHb is a homodimer oxygen-binding protein frequently used to treat hypoxia in microorganism cultures. The findings imply that under conditions of high oxygen tension, *G. xylinus* prefers cell expansion rather than BC synthesis. Instead, lower oxygen tension encourages the formation of BC, and when oxygen tension is low, VHb-positive *G. xylinus* produces noticeably more BC than *G. xylinus* without VHb (Liu, Li, Xie, et al., 2018).

7.2.5 Long-Term Fermentation

A waiting period (lag phase) for the starting time is unavoidable during the batch fermentation process, and this waiting period is unproductive in terms of economic viability. Undiluted molasses is a better culture medium when compared to HS and Zhou medium, according to a study using the batch-fed culturing technique (Çakar, Ozer, Aytekin, et al., 2014). Additionally, the semicontinuous approach used in the experiment resulted in a longer BC production period of up to 28 days. Additionally, extensive study on PCS-RDB demonstrates the potential for semicontinuous, long-term fermentation, as previously mentioned (Lin, Hsieh, Chen, et al., 2014).

7.2.5.1 Other Additives for Efficient BC Production

To either boost production or enhance the mechanical qualities of BC, or to do both, additives were added to the media. The culture medium was supplemented with water-soluble PVA to investigate the in situ modification impact. In comparison to the native BC, this PVA-reinforced BC had improved mechanical properties. Young's modulus and tensile strength at the break were increased by 15, 165, 680% and 1, 12, 40%, respectively, with additions of 0.6, 6 and 14 wt% of PVA (Castro, Vesterinen, Zuluaga, et al., 2014). For in situ modification, the PCS-RDB technique is used with other chemicals. The synthesis of BC increased by 80% and 113% when carboxymethylcellulose (CMC), avicel, agar, and sodium alginate were added to the medium. Further investigation, however, revealed that these additives—which are absorbed into the BC fibrils throughout the fermentation process—decreased the total crystallinity of the final BC produced (Lin, Liu, Hsu, et al., 2016). Ethanol and acetic acid have been investigated for their potential to increase BC productivity (Cacicedo, Castro, Servetas, et al., 2016). As alternate energy sources, ethanol and acetic acid were recently introduced to the modified HS medium. The outcomes showed that the addition of ethanol and acetic acid stimulates BC synthesis. Comparing a medium with and without these two chemicals, 0.1 wt% of ethanol and acetic acid can boost BC synthesis by up to 279% and 222%, respectively (Molina-Ramirez, Enciso, Torres-Taborda, et al., 2018). Ascorbic acid, also referred to as vitamin C, is regarded as a powerful antioxidant. When included in the medium at 0.5% w/w, it has been found to increase BC production by 188% and decrease the synthesis of gluconic acid (Keshk, 2014). The impact of ethylene, a well-known phytohormone that controls plant development, on BC production is also being researched. *BcsA* and *BcsB* are upregulated directly by the presence of ethylene (in situ) generated by ethephon, and *cmcAx, ccpAx,* and *bglAx,* another group of genes known for their function in BC synthesis, are indirectly regulated. Indole-3-acetic acid (IAA), like other phytohormones, reduces BC synthesis by downregulating *bcsA* expression (Augimeri & Strap, 2015).

7.2.6 Modification of Bacterial Cellulose

Although BC has numerous advantageous characteristics for a variety of applications, its spectrum of usage still requires modification and the addition of additional molecules. To improve BC's mechanical qualities, various physical processes can be used. It has been shown that increasing the developing polymer chain's exposure to a rotating magnetic field increases water absorption and BC density (Fijalkowski, Zywicka, Drozd, et al., 2017). After analyzing the rheology of BC during ultrasound treatments, it was shown that brief treatments of polymer suspensions for 1 min caused the fibers to break down into half of their original width, enhancing both the material's stability (Paximada, Dimitrakopoulou, Tsouko, et al., 2016) and water holding capacity. Such changes could improve BC's qualities for use as a food additive or as a material for highly exuding wound dressing in the biomedical industry. BC's chemical alterations have lately undergone a thorough evaluation (Habibi, 2014).

The oxidation of cellulose's hydroxyl groups is a highly frequent alteration, and 2,2,6,6-tetramethyl-1-piperidinyloxyl (TEMPO) is frequently used to do it. TEMPO particularly targets primary hydroxyls to produce carboxylated cellulose (CBC). When Ag is assembled on CBC surfaces, these carboxyl groups can be employed to covalently link functional molecules, such as photoluminescent particles after amine activation of the CBC, or to change the surface charge for better ionic interactions. BC can also be acetylated or given an aminoalkyl graft as additional chemical changes. However, acid hydrolysis of BC for the creation of nanocrystals is one of the most common chemical alterations. Additionally, BC has been employed in a number of composite compositions. By using in situ techniques, BC can be blended with other polymers. To do this, simply incorporate the composite polymer into the culture media. By doing this, it is possible to create a hyaluronic acid-BC composite with a rougher structure and more hydrophobic BC chains. Additionally, the addition of $CaCO_3$ led to increased O/C ratios, which gave the composite its amphoteric surface property. Ex situ synthesis can also be used to create BC-polymer composites. Chitosan (CS)-BC composite, created via straightforward solvent dissolving and casting, and CS/PVP/BC manufactured using salt leaching procedures are two examples of this practice. Recently reviewed BC composites with metal-oxide and oxide nanoparticles often entail strong mixing of BC and the metallic nanoparticles. Immersion into a water solution containing the desired ingredient has remained the predominant method of introducing active compounds into BC matrices up to this point.

7.2.7 Parameters Influencing Production of Bacterial Cellulose

To increase BC yield, it is necessary to optimize a number of variables, including temperature, pH, and dissolved oxygen, regardless of whether the cultivation is static or agitated. Microstructure is influenced by cellulose I, size, crystalline polymorph, and crystallinity index (CI), all of which depend on the growing circumstances (Yan, Chen, Wang et al., 2008).

7.2.7.1 Temperature

Temperature is one of the most crucial factors since it can control how an organism adapts to survive by affecting its typical homeostatic physiology. As demonstrated by a *Komagataeibacter* sp. cultivation at 30°C for 7 days under static conditions (Zahan, Hedzir, & Mustapha, 2017), a temperature range of 25°C to 30°C was determined to be ideal for the production of BC, whereas *Acetobacter xylinum* (Aswini, Gopal, & Uthandi, 2020) prefers a temperature of about 28°C. *Acetobacter senegalensis* MA1 required a slightly higher temperature of 33.5°C (Rangaswamy, Vanitha, & Hungund,

2015). The optimal temperature range for the generation of BC by *Gluconacetobacter* sp. RV28, *Pseudomonas* sp. RV14, and *Enterobacter* sp. RV11 was discovered to be 28°C to 30° C, as high temperatures promote denaturation of the culture (Rangaswamy, Vanitha, & Hungund, 2015).

7.2.7.2 Culture Media

The primary elements of the growth medium required for BC fermentation are carbon, in the forms of fructose, glycerol, maltose, starch, and xylose, and nitrogen, in the forms of casein hydrolysate and peptone. Changes to the growth media might be either direct or indirect. Effects on the growth patterns of microbes, a yield of 5 g/L of BC and a yield of 4.8 g/L of BC were reported in the presence of casein hydrolysate and peptone, respectively, from *A. xylinum*. Vitamins have a crucial role in controlling cellular growth and metabolism. In addition to riboflavin and pantothenate, as well as the vitamins pyridoxine, nicotinic acid, and biotin, cell production requires *p*-aminobenzoic acid. A low rate of gluconic acid production was found to boost the productivity of lignosulfonate-containing BC.

7.2.7.3 pH

Another crucial element in regulating the oxidative fermentation of BC synthesis is pH.

BC manufacturing is ideal in acidic or nearly neutral pH conditions. The pH of the fermentation culture medium is altered during BC fermentation as a result of the generation of secondary metabolites such as acetic acid, gluconic acid, and lactic acid. As a result, pH 4 to 6 is thought to be the optimal range for BC fermentation culture medium. *Acetobacter xylinum,* another strain of *Acetobacter xylinum ie Acetobacter sp. A9,* and *Komagataeibacter* spp. require pH of 5.50, 4.5 to 7.5, and 6.0, respectively, according to experimental findings.

7.2.7.4 Agitation Rate

Another crucial factor, agitation rate, influences the production of BC. A solid ball with a diameter of 0.5 to 1 cm forms with a lower agitation rate of 100 rpm.

BC size diminishes as agitation rate rises between 150 and 250 rpm, and at 300 rpm, clumps of irregular shape start to form.

7.2.7.5 Growth Curve

The growth curve patterns of BC-producing bacterial strains are the same as those of the other bacterial strains, including lag, log, stationary, and death phases. When *Komagataeibacter mendellinensis* is grown in media containing 2% w/v of glucose, fructose, and sucrose as carbon sources, the lag phase lasts for 25 hours. In the case of *G. liquefaciens,* 5% fructose in the culture medium lengthens the lag phase by about 10 to 15 hours, resulting in the prolonged consumption in BC synthesis.

7.2.7.6 O_2 level

Oxygen is crucial to control aeration in the growth medium. Since all the bacteria in the culture medium are aerobic in nature, there needs to be a sufficient supply of oxygen. Low levels of dissolved oxygen in the medium prevent bacterial growth, which slows the formation of BC. The generation of BC and cellular metabolism both depend on oxygen. In addition to causing the BC output to collapse, a limited O_2 supply also lowers the BC quality. The viscosity of the broth is inversely correlated with the O_2 transfer rate. The O_2 transfer rate and BC productivity both decline as viscosity rises.

7.2.7.7 Yield

Independent of the method of cultivation, different bacterial strains prefer different nutritional media. The production of BC also benefits from the use of inexpensive materials such molasses, pineapple peel, and citrus waste solution. The best time to control BC cultivation to get the best output is between 2 and 14 days. A general increase in incubation days could boost BC productivity. Among all other sources, glucose is the primary nutritional source of carbon that bacteria use extensively to meet their metabolic needs and produce large amounts of energy.

7.3 STRUCTURE OF BACTERIAL CELLULOSE

BC is a kind of fibril that has as its fundamental structural component a β-1→4 glucan chain with the chemical formula $(C_6H_{10}O_5)n$. Inter- and intrahydrogen bonds keep the glucan chains together. BC microfibrils are around 100 times smaller than those of plant cellulose and were initially identified by Muhlethaler in 1949.

A hydrogel sheet with a high surface area and porosity is created by the well-arranged three-dimensional nanofibers in the BC's fibrous network. *Acetobacter xylinum* generates the thermodynamically stable polymer cellulose II and the ribbon-like polymer cellulose I. Protofibrils of the glucose chain are released through the cell wall of the bacterium throughout the production process, where they combine to create nanofibrils of cellulose ribbons. With a very porous matrix, these ribbons provide the web-shaped network structure of BC. According to Klemm et al. (2005), the cellulose produced has a large number of hydroxyl groups on its surface, which accounts for its hydrophilicity, biodegradability, and chemical modification ability.

7.4 PROPERTIES OF BACTERIAL CELLULOSE

The most widely used and sustainable polymer is cellulose, which is also the most versatile. Although cellulose is generally thought of as a plant-based material, numerous fermentation techniques have produced alternate sources of cellulose from bacterial genera with promising results. Despite having diverse physical and chemical characteristics, BC and plant fiber share a similar structural makeup, with two distinct cellulose subunits, I and II. The degree of cellulose polymerization ranges from 2000–6000 in BC to 13,000–14,000 in plant cellulose. The cellulose's three-dimensional mesh-like structure shields it against microbial attack. The free surface hydroxyl group on the cellulose's surface is a result of the β-1 → 4 glycosidic bonds. BC has a diameter of 10–50 nm and a length of 100–1000 nm.

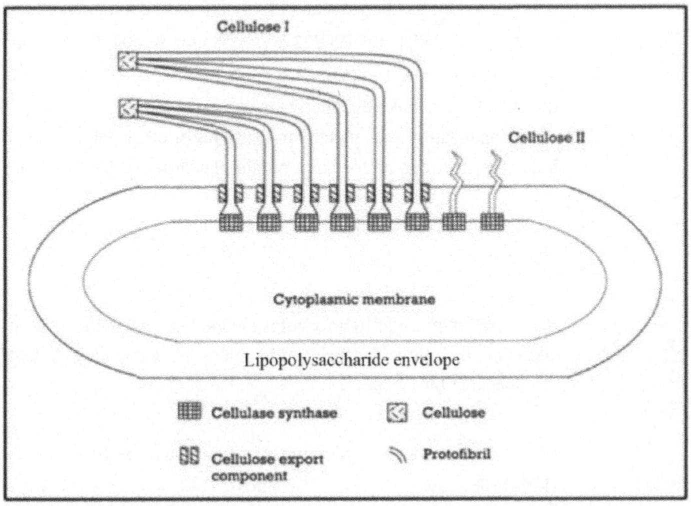

FIGURE 7.3 Production of cellulose microfibrils by *Acetobacter xylinum*.

7.5 APPLICATION OF BACTERIAL CELLULOSE FOR SUSTAINABLE PRODUCTION AND CONSUMPTION

One of the top priorities for microbial metabolic engineers is sustainable development. The Brundtland Report, issued in 1987 by the United Nations World Commission on Environment and Development, provided the first definition of the modern idea of sustainable development. The goal of the report was to provide a solution that promotes economic growth while also enabling environmental conservation. Through subsequent international initiatives, such as Agenda 21 in 1992 and the Millennium Development Goals (MDGs) in 2000, both led by the United Nations, the meaning of sustainable development has continued to change. The United Nations General Assembly officially adopted a new set of seventeen quantifiable Sustainable Development Goals (SDGs) in September 2015. These goals range from eradicating global poverty and hunger to combating climate change by 2030.

TABLE 7.3
Brief Description of Applications of Bacterial Cellulose (BC)

Pharmaceutical Application	
Wound dressing	BC is a suitable material for wound healing on its own or when reinforced with charged molecules due to its water holding capacity, biocompatibility, and mechanical qualities.
Antimicrobial activity	A component that prevents infection can be created by functionalizing antimicrobial molecules or combining them with other elements in composites.
Drug delivery	BC is emerging as a viable platform for drug administration due to its propensity for drug adsorption and retaining water, both with regulated release profiles.
Tissue regeneration	BC is an effective scaffold material that can be further improved physically or chemically. It has been employed as a scaffold for the in vitro development of several cell types.
Vascular grafts	BC is an excellent material for artificial blood arteries due to its water holding capacity, flexibility, biocompatibility, and mild inflammatory response.
Food Applications	
Emulsion stabilizer	BC nanoparticles act as an oil-in-water emulsion stabilizer.
Gastric conditions	BC fibers prevent constipation.
Filtration	For specialized components and highly effective contaminant removal, BC can be utilized as a matrix for coating with other carbon-based or metal compounds.
Electronic Applications	
Conductive material	It can be turned into a conductive material for the use of nanowires or batteries after functionalization with metal nanoparticles or other polymers.
Sensors	BC nano paper can be used as a sensor platform for chemicals or bioreactions.
Surface Enhanced Raman Spectroscopy (SERS)	BC functionalized with nanoparticles has favorable properties as a platform for SERS spectroscopy.
Energy Production	
Bioethanol	BC is used in place of lignocellulose made from plants. Hemicellulosic and lignin-deficient BC allows for the elimination of the acid-catalyzed hydrolysis phase in the synthesis of bioethanol.
Miscellaneous	
Cosmetics	BC facial masks are moisturizing, and functionalized BC can address specific cosmetic problems.
Paper manufacturing	BC as an additive increases material strength.
Aerogels	BC as the primary polymer, composited with other biocompatible polymers.

In this aspect, the goal of metabolic engineering and the SDGs, particularly SDG 12's mandate to "ensure sustainable consumption," are closely aligned and patterns of manufacturing. synthesis of chemicals and minerals using microorganisms and renewable resources. SDG 12 resources can benefit the environment and the economy. Metabolic systems engineering, which incorporates synthetic biology, evolutionary engineering, and metabolic engineering, has made it possible to produce an expanding number of chemicals and materials using biobased methods. BC has a great deal of potential to revolutionize how we live and run enterprise. Since BC's contribution to the environment through its biodegradability is fairly evident (e.g., by utilizing cellulase), new economic opportunities exist for the use of BCs in industries.

The physical characteristics of BC are superior to those of plant cellulose in terms of mechanical stability, tensile strength, thermostability, crystallinity (i.e., ultrafine fiber structure), purity, and biocompatibility (i.e., biologically nontoxic and immunological responses are absent). Excellent biocompatibility of BC has encouraged the development of BC-based biomedical goods such as bone tissue scaffolds, artificial blood vessels, artificial skin, and dental implants, even though they have not yet been fully marketed. In tissue engineering and synthetic scaffolds, cell adhesion to the BC surface is crucial. As a result, numerous chemicals that encourage cell attachment have been tried. Fermentation produces high-purity BC that makes separation, purification, and additional physical and chemical alterations easier. The application range of BC has been widened from the food business to the cosmetics, biomedical, and even electronics industries because of its competitive advantages (Table 7.3). It is interesting to note that BC is expanding into sectors like the battery and display industries that appear to have nothing to do with biology. Demand for energy storage is rising as more electric vehicles and portable devices are developed. Because of its high tensile strength, high thermostability, and high crystallinity, BC can be used as a battery separator in electrochemical energy storage systems, such as Li-ion batteries, as one outstanding example of potential new uses for BC. For the battery separator to effectively transport ions while preventing direct contact between the anode and cathode, it must be physically and chemically stable. The battery's performance is greatly influenced by its structure and qualities. BC is regarded as an excellent contender for the next-generation battery separator due to its distinctive fibrous and cross-linked three-dimensional network structure. Even though we only used BC as one example of how it aids in "sustainable production and consumption" here, many other biobased goods would also support this SDG. To achieve sustainable consumption and production, more research is needed to solve outstanding issues, such as low production levels and material qualities that can still be improved.

7.6 CONCLUSIONS

Through oxidative fermentation, acetic acid bacteria are able to take the place of plants as an alternate source of cellulose in static, agitated, and stirring environments. However, its commercial applications and industry potential are constrained by high production costs and limited yields. Consequently, studies have concentrated on agriculture waste and industrial waste as new, affordable carbon sources. In addition, the revision and inclusion of particles in the BC matrix improve its physiological and natural qualities and present a new potential for applications.

7.7 REFERENCES

Aswini K, Gopal NO, Uthandi S. Optimized culture conditions for bacterial cellulose production by Acetobacter senegalensis MA1. BMC Biotechnol. 2020;20:46.

Augimeri RV, Strap JL. The phytohormone ethylene enhances cellulose production, regulates CRP/FNRKx transcription and causes differential gene expression within the bacterial cellulose synthesis operon of Komagataeibacter (Gluconacetobacter) xylinus ATCC 53582. Front Microbiol. 2015;6:1459.

Aydin YA, Aksoy ND. Isolation and characterization of an efficient bacterial cellulose producer strain in agitated culture: *Gluconacetobacter hansenii* P2A. Appl Microbiol Biotechnol. 2014;98(3):1065–1075.

Blanco Parte FG, Santoso SP, Chou CC, et al. Current progress on the production, modification, and applications of bacterial cellulose. Crit Rev Biotechnol. 2020 January 14;40(3):397–414. https://doi.org/10.1080/07388551.2020.1713721

Cacicedo ML, Castro MC, Servetas I, et al. Progress in bacterial cellulose matrices for biotechnological applications. Bioresour Technol. 2016;213:172–180.

Çakar F, Ozer I, Aytekin AO, et al. Improvement production of bacterial cellulose by semi-continuous process in molasses medium. Carbohydr Polym. 2014;106:7–13.

Castro C, Vesterinen A, Zuluaga R, et al. In situ production of nanocomposites of poly(vinyl alcohol) and cellulose nanofibrils from Gluconacetobacter bacteria: Effect of chemical crosslinking. Cellulose. 2014;21(3):1745–1756.

Cavka A, Guo X, Tang S-J, et al. Production of bacterial cellulose and enzyme from waste fiber sludge. Biotechnol Biofuels. 2013;6(1):25.

Cheng K-C, Catchmark JM, Demirci A. Enhanced production of bacterial cellulose by using a biofilm reactor and its material property analysis. J Biol Eng. 2009;3(1):12.

Fan X, Gao Y, He W, et al. Production of nano bacterial cellulose from beverage industrial waste of citrus peel and pomace using Komagataeibacter xylinus. Carbohydr Polym. 2016;151:1068–1072.

Fijalkowski K, Zywicka A, Drozd R, et al. Increased water content in bacterial cellulose synthesized under rotating magnetic fields. Electromagn Biol Med. 2017;36(2):192–201.

Gorgieva, Trček. Bacterial cellulose: Production, modification and perspectives in biomedical applications. Nanomaterials. 2019 September 20;9(10):1352. https://doi.org/10.3390/nano9101352

Habibi Y. Key advances in the chemical modification of nanocelluloses. Chem Soc Rev. 2014;43(5):1519–1542.

Hommel R. Acetobacter. In C. A. Batt & M. Lou Tortorello (Eds.), *Encyclopedia of Food Microbiology*. 2014, 2nd ed., 3–10. Elsevier. https://doi.org/10.1016/b978-0-12-384730-0.00001-x

Huang C, Guo H-J, Xiong L, et al. Using wastewater after lipid fermentation as substrate for bacterial cellulose production by Gluconacetobacter xylinus. Carbohydr Polym. 2016;136:198–202.

Jozala AF, Pertile RAN, dos Santos CA, et al. Bacterial cellulose production by Gluconacetobacter xylinus by employing alternative culture media. Appl Microbiol Biotechnol. 2015;99(3):1181–1190.

Keshk SM. Vitamin C enhances bacterial cellulose production in Gluconacetobacter xylinus. Carbohydr Polym. 2014;99:98–100.

Klemm D, Heublein B, Fink HP, Bohn A. Cellulose: fascinating biopolymer and sustainable raw material. Angew Chem Int Ed Engl. 2005;44(22):3358–3393. doi:10.1002/anie.200460587

Kuo C-H, Chen J-H, Liou B-K, et al. Utilization of acetate buffer to improve bacterial cellulose production by Gluconacetobacter xylinus. Food Hydrocoll. 2016;53:98–103.

Li Z, Wang L, Hua J, et al. Production of nano bacterial cellulose from waste water of candied jujubeprocessing industry using Acetobacter xylinum. Carbohydr Polym. 2015;120:115–119.

Lin D, Lopez-Sanchez P, Li R, et al. Production of bacterial cellulose by Gluconacetobacter hansenii CGMCC 3917 using only waste beer yeast as nutrient source. Bioresour Technol. 2014;151:113–119.

Lin S-P, Hsieh S-C, Chen K-I, et al. Semi-continuous bacterial cellulose production in a rotating disk bioreactor and its materials properties analysis. Cellulose. 2014;21(1):835–844.

Lin S-P, Huang Y-H, Hsu K-D, et al. Isolation and identification of cellulose-producing strain Komagataeibacter intermedius from fermented fruit juice. Carbohydr Polym. 2016;151:827–833.

Lin S-P, Liu C-T, Hsu K-D, et al. Production of bacterial cellulose with various additives in a PCS rotating disk bioreactor and its material property analysis. Cellulose. 2016;23(1):367–377.

Lin S-P, Loira Calvar I, Catchmark JM, et al. Biosynthesis, production and applications of bacterial cellulose. Cellulose. 2013;20(5):2191–2219.

Liu M, Li S, Xie Y, et al. Enhanced bacterial cellulose production by Gluconacetobacter xylinus via expression of Vitreoscilla hemoglobin and oxygen tension regulation. Appl Microbiol Biotechnol. 2018;102(3):1155–1165.

Liu M, Zhong C, Wu X-Y, et al. Metabolomic profiling coupled with metabolic network reveals differences in Gluconacetobacter xylinus from static and agitated cultures. Biochem Eng J. 2015;101:85–98.

Matsutani M, Ito K, Azuma Y, et al. Adaptive mutation related to cellulose producibility in Komagataeibacter medellinensis (Gluconacetobacter xylinus) NBRC 3288. Appl Microbiol Biotechnol. 2015;99(17):7229–7240.

Mohite BV, Salunke BK, Patil SV. Enhanced production of bacterial cellulose by using Gluconacetobacter hansenii NCIM 2529 strain under shaking conditions. Appl Biochem Biotechnol. 2013;169(5):1497–1511.

Molina-Ramırez C, Castro M, Osorio M, et al. Effect of different carbon sources on bacterial nanocellulose production and structure using the low pH resistant strain Komagataeibacter medellinensis. Materials. 2017;10(6):639.

Molina-Ramirez C, Enciso C, Torres-Taborda M, et al. Effects of alternative energy sources on bacterial cellulose characteristics produced by Komagataeibacter medellinensis. Int J Biol Macromol. 2018;117:735–741.

Moniri M, Boroumand Moghaddam A, Azizi S, et al. Production and status of bacterial cellulose in biomedical engineering. Nanomaterials (Basel). 2017;7(9):257.

Morgan JL, Strumillo J, Zimmer J. Crystallographic snapshot of cellulose synthesis and membrane translocation. Nature. 2013;493(7431):181–186.

Park JK, Park YH, Jung JY. Production of bacterial cellulose by *Gluconacetobacter hansenii* PJK isolated from rotten apple. Biotechnol Bioprocess Eng. 2003;8(2):83–88.

Paximada P, Dimitrakopoulou EA, Tsouko E, et al. Structural modification of bacterial cellulose fibrils under ultrasonic irradiation. Carbohydr Polym. 2016;150:5–12.

Rangaswamy BE, Vanitha KP, Hungund BS. Microbial cellulose production from bacteria isolated from rotten fruit. Int. J. Polym. Sci. 2015;2015:280784.

Ross P, Mayer R, Benziman M. Cellulose biosynthesis and function in bacteria. Microbiological Reviews. 1991 March;55(1):35–58. https://doi.org/10.1128/mr.55.1.35-58.1991

Saichana N, Matsushita K, Adachi O, et al. Acetic acid bacteria: A group of bacteria with versatile biotechnological applications. Biotechnol Adv. 2015;33(6):1260–1271.

Singhsa P, Narain R, Manuspiya H. Physical structure variations of bacterial cellulose produced by different Komagataeibacter xylinus strains and carbon sources in static and agitated conditions. Cellulose. 2018;25(3):1571–1581.

Sun Run-Cang. Detoxification of biomass for bioethanol. BioResources. 2008;4(2):452–455.

Tsouko E, Kourmentza C, Ladakis D, et al. Bacterial cellulose production from industrial waste and byproduct streams. Int J Mol Sci. 2015;16(12):14832–14849.

Urbina L, Hernandez-Arriaga AM, Eceiza A, et al. Byproducts of the cider production: An alternative source of nutrients to produce bacterial cellulose. Cellulose. 2017;24(5):2071–2082.

Uzyol HK, Sacan MT. Bacterial cellulose production by Komagataeibacter hansenii using algae-based glucose. Environ Sci Pollut Res Int. 2017;24(12):11154–11162.

Wu J-M, Liu R-H. Thin stillage supplementation greatly enhances bacterial cellulose production by Gluconacetobacter xylinus. Carbohydr Polym. 2012;90(1):116–121.

Wu S-C, Li M-H. Production of bacterial cellulose membranes in a modified airlift bioreactor by Gluconacetobacter xylinus. J Biosci Bioeng. 2015;120(4):444–449.

Yan Z, Chen S, Wang H, et al. Biosynthesis of bacterial cellulose/multi-walled carbon nanotubes in agitated culture. Carbohydr. Polym. 2008;74:659–665.

Yin N, Santos TMA, Auer GK, et al. Bacterial cellulose as a substrate for microbial cell culture. Appl Environ Microbiol. 2014;80(6):1926–1932.

Zahan KA, Hedzir MSA, Mustapha M. The potential use of papaya juice as fermentation medium for bacterial cellulose production by Acetobacter xylinum 0416. Pertanika J. Trop. Agric. Sci. 2017;40:343–350.

Zhong C. Industrial-scale production and applications of bacterial cellulose. Front Bioeng Biotechnol. 2020 December 22;8. https://doi.org/10.3389/fbioe.2020.605374

8 Preparation of Bacterial Cellulose Composites by *in Situ* Processes

Amitabh Aharwar
Government College, Harrai
Chhindwara, Madhya Pradesh, India

Archana Bhaskar
Rajiv Gandhi Arts and Commerce College
Mungeli, Chhattisgarh, India

Dharmendra Kumar Parihar
Guru Ghasidas Vishwavidyalaya
Bilaspur, Chhattisgarh, India

Anil Kumar Patel and Reeta Rani Singhania
National Kaohsiung University of Science and Technology
Kaohsiung, Taiwan

8.1 INTRODUCTION

The most prevalent renewable polymer resource on earth is cellulose, a polymer made of β linked glucose molecules that are produced by a wide range of organisms such as plants, algae, bacteria, and certain animal species as well. Numerous businesses, including the food and paper industries, the pharmaceutical industry, and the medical field, employ cellulose (Volova et al., 2022). Plant cellulose has extensive uses in many different sectors, yet it is thought to have some severe flaws. The most significant of these is its impure character and the requirement for rigorous pretreatments before it can be used. The unbranched polymer known as bacterial cellulose (BC) is made up of linear chains of -1,4-glucopyranose residues (Eslahi et al., 2020).

Gluconacetobacter (or *Komagataeibacter*), *Acanthamoeba, Achromobacter, Zooglea, Agrobacterium, Aerobacter, Azotobacter, Rhizobium, Sarcina, Salmonella, Escherichia, Pseudomonas,* and *Alcaligenes* are among the taxa that produce BC extracellularly. A. J. Brown identified BC in vinegar during fermentation in 1886 (Rajwade et al., 2015). BC is a refined form of cellulose created through fermentation by a variety of bacterial species, primarily those from the genera *Acetobacter* and *Komagataeibacter* (Fernandes and Gama, 2019). Bacterial cellulose is created as a gelatinous film made of pure cellulose fibers arranged in a three-dimensional (3D) nanofibrillar pattern. A 3D web-shaped arrangement of BC creates a porous geometry with remarkable mechanical stability. BC is a perfect biomaterial for many uses in the medical and other industries since it has substantially higher crystalline structure, liquid absorption capacity, and mechanical qualities than plant cellulose (Ul-Islam et al., 2015). Bacterial cellulose is about 1000 times thinner than conventional plant cellulose fibrils. BC is a 3D network of nano- and microfibrils with 3–4 nm thickness and 70–80 nm length in dimensions. Hemicelluloses, lignin, and other substances

 DOI: 10.1201/9781003355434-8

found naturally typically connected to cellulose extracted from plant cell walls are absent from BC. Because of its randomly arranged 3D network of interconnected nanofibers with a diameter of 20–100 nm and several micrometers in length, BC has unique mechanical properties, such as a high level of polymerization, high water holding capacity, greater moldability, high tensile strength, high surface area, and crystallinity (Tseng et al., 2022). These characteristics are very helpful for the creation of composite materials (Dluga et al., 2021; Wu et al., 2016).

Bacterial cellulose is synthesized by *Gluconacetobacter* sp. from a specific substrate (glucose) through two metabolic pathways, the phosphate pentose pathway, which allows the oxidation of carbohydrates, and the Krebs cycle, which allows the oxidation of fat acids. Overall, five steps make up the fundamental process for the BC biosynthesis pathway: glucokinase phosphorylates glucose to glucose-6-phosphate; phosphoglucose mutase isomerizes glucose-6-phosphate to glucose-1-phosphate; uridine diphosphate glucose (UDPG) pyrophosphorylase converts glucose-1-phosphate to UDPG; cellulose synthase produces linear -1,4-glucan chains from UDPG; and BC is crystallized from the polymerized subunit (Singhania et al., 2021) (Figure 8.1). BC molecules are produced within the bacterium during the biosynthesis process, and then they are released through a row of holes (squeezing points) on the longitudinal axis of the bacterial cell's (Liu et al., 2020).

Bacterial cellulose works effectively as a reinforcing and supporting component in nanocomposites made of a variety of polymeric matrices, including plasticized starch, epoxidized soybean oil matrix, polylactic acid (PLA), phenolic resins, acrylic thermosetting resins, and cellulose acetate butyrate. Bacterial cellulose is currently used industrially and commercially for a variety of products, from foods to functional substances like the diaphragms in speakers and headphones, electronic devices, wound dressings, high-quality paper additives, artificial skin, cellulose derivatives, optic material, edible strips, and membrane filters. The use of BC extends to a variety of industries, including the paper industry, packaging, and biosensors. Such uses have been made possible by either employing BC in its natural state or combining it with other organic substances, biopolymers, and inorganic compounds, such as metal nanoparticles and nanowires. Because there are so many hydroxyl groups on the surface, composites of BC can be made. Both organic and inorganic substances have been investigated during fermentative synthesis, and it has been demonstrated that these materials incorporate into the fibrillar network as BC is being made by the bacteria. As a result, the latter kinds of composites might not be uniform (Rajwade et al., 2015; Volova et al., 2022).

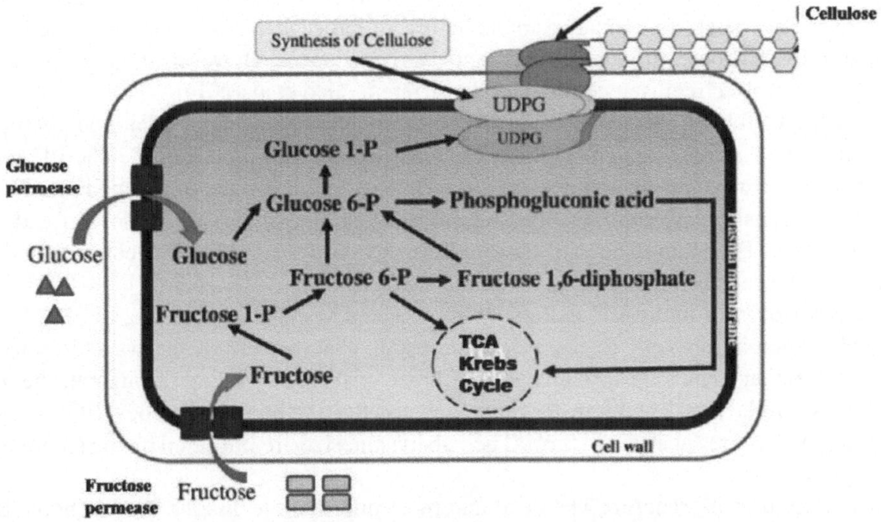

FIGURE 8.1 Synthesis of bacterial cellulose.

Technology using polymer composites has become extremely well known. Although composites enhance the already-existing capabilities of polymers, pure polymers nevertheless have limitations. To improve the properties and applications of BC composites, a variety of processes are used to produce the materials. Web-shaped microfibrils that can trap different materials found in culture medium throughout the fabrication process are also capable of allowing material adhesion towards the microfibril (Shah et al., 2013). The antibacterial and biocompatible properties of BC have been improved by impregnation with nanomaterials with antibacterial properties, such as silver (Ag), zinc oxide (ZnO), titanium oxide (TiO2), antibacterial clays, and some polymers. As a result, its applications in medical fields have increased. Parallel to how BC gained conducting properties with the inclusion of conducting materials, such as carbon nanotubes (CNTs), graphene oxide, PEDOT:PSS, and nanoparticles. The fermentation process and bacterial activity have a significant impact on productivity (Ul-Islam et al., 2015; Foresti et al., 2017).

This chapter deals with bacterial cellulose production and bacterial cellulose composite preparation. The in situ method of bacterial cellulose composite preparation is explained and different cases are discussed.

8.2 BACTERIAL CELLULOSE PRODUCTION

Microbes produce BC to hold onto moisture and stop the bacteria from drying up. It aids in the bacteria's ability to float by trapping carbon dioxide created during the tricarboxylic acid cycle. By doing so, the organism is kept in an aerobic environment. Due to its impenetrable nature, the cellulose pellicle shields the bacterium from the harmful effects of ultraviolet radiation. Additionally, it also protects bacteria mechanically, chemically, and physiologically. In symbiotic or infectious relationships, bacterial cellulose encourages cell attachment, shields the bacterium from heavy metal ions, and enhances the flow of nutrients. In bacteria that are connected with plants, cellulose biofilm aids in the bacteria's apposition to plant tissue, giving it a survival advantage naturally (Gabryś et al., 2022; Rajwade et al., 2015).

Bacterial cellulose is mostly produced in bioreactors, shaking or agitated cultures, and static cultures. The main criteria for choosing manufacturing techniques are the final uses, mechanical, physical, and morphological characteristics because each technique produces a product with a distinct microstructure, macroscopic appearance, and characteristics. Pellicles occur between the air-liquid interfaces in the culture medium whenever the static culture method is used, whereas irregular masses that resemble pellets, spheres, or asterisks emerge when the culture is agitated. Different uses necessitate fermentative manufacture in either static, agitated, or stirred settings. Traditionally, BC has been produced under static circumstances. However, the types of cellulose generated under these circumstances are wholly different (Gao et al., 2019).

Static cultures for BC film production: Bacterial cellulose films or gelatinous pellicles are products of static cultures, the approach most extensively used to produce BC. The BC forms on the liquid's surface or at the air-liquid interface and floats to the surface of the medium. The static culture approach has the lowest production rate and is the slowest process, but it does provide BC for many applications. The main issue with static culture techniques is the insufficient oxygen delivery, particularly to the cells below the BC layer on the surface.

Shaking cultures to make BC pellets: When created via shaking culture, the BC formed is considerably different in shape. Because of the constant shaking, the BC grows as tiny irregular-shaped pellets rather than a consistent coating. Constant oxygen blending throughout the medium encourages BC production. Therefore, this approach was the first one used to boost BC biosynthesis. The resulting pellets are spherical in form. Their shape and size are impacted by the spinning speed (Ul-Islam et al., 2015).

BC production in bioreactor: The third and most important technique used to boost BC output was high-speed agitation cultures in specially designed reactors or fermenters, in which quick nutrient or growth media input and oxygen blending are guaranteed. In this culture method, BC is generated as little granules. Because of the shear stress created by agitation, however, agitated cultures produce fibrous BC with a lower degree of polymerization, a lower crystallinity index, and a lower Young's modulus. To create cellulose-based composites, several inert substances have been introduced to fermentative environments. These materials have been investigated for both medical and nonmedical purposes. Agro-wastes and other inexpensive, widely accessible substrates have all been tried out the fermentative synthesis of BC. Fruit juices, sugarcane molasses, corn steep liquor, and wastes from the wood processing industries have also been evaluated in addition to the main ones as substrates for the production of BC (Rajwade et al., 2015). The accumulation of BC in shaking cultures causes the medium's viscosity to rise and the culture's homogeneity to fall. This lowers the oxygen content in the medium, which has an impact on bacterial growth and cellulose production (Ul-Islam et al., 2015).

8.3 BACTERIAL CELLULOSE COMPOSITES

In 1986 composites were first defined by Roy et al. (1986) as a compound composed of several phases in which one of the phases was a nanometric scale length (Roy et al., 1986). The matrix and the reinforcing material are two different types of composites. The reinforcements add physiochemical and biological characteristics to the matrix, which serves as a scaffold and holds the reinforcement material. A variety of composite materials with optimum qualities and uses can be synthesized using a wide range of matrices and reinforcement components. Different types of reinforcing elements, such as fibrous materials, metallic oxides, solid clay particles, and carbonates, have strengthened polymers while also adding a variety of biological properties to polymeric composites (Shah et al., 2013). Both a matrix and a reinforcing material can be made of cellulose. The cellulose's well-organized fibrous network system can cage in nanopolymers, serving as a matrix. Cells, tissues, and other polymeric and nonpolymeric materials can also be strengthened by cellulose nanofibers. Due to its identical chemical and superior physical characteristics to plant cellulose, BC has been used to create a variety of composite materials for use in a variety of applications (Wu et al., 2016, Shah et al., 2013).

Various approaches to the synthesis of BC-based composites have been used, depending on the intended use. The three techniques that are most frequently used to prepare BC composites are the in situ strategy, ex situ strategy, and the synthesis of BC composite from BC solution. The in situ method incorporates reinforcing materials into the culture medium during BC production that ultimately end up in the BC hydrogel created. The ex situ process involves inserting or impregnating reinforcing components into a synthetic polymer to create composites. Based on the developing strategy and the way that materials are combined, this approach can be further specified. For BC-based composites, the synthesis procedure is much more in controlled because it is accomplished by using a dissolving BC synthesizing solution (Shah et al., 2013). The composition of the BC matrix and reinforcement agents can be controlled via the synthesis process.

Due to their outstanding qualities, composite materials, both natural and manmade, are currently utilized for substantially all industrialized procedures. In composites, a matrix serves as the foundation and a scaffold, and reinforcement components provide the matrix additional properties. Numerous composites can be created using a wide range of matrices and reinforcing components using a variety of synthetic processes. To create multifunctional composites, polymers, including biopolymers, have been filled with a variety of reinforcing components. BC is a superior biopolymer for the creation of composite materials. The creation of BC composites is primarily driven by two factors. First, the exceptional structural characteristics of BC make it a perfect candidate for the synthesis of composites made of other materials. Nanoparticles can be contained by the beautifully organized web-shaped fibrous network structure of BC fibers, acting as a matrix. In addition, the hydrogen-bonded sites on the cellulose chains allow for interactions with other polymers to create robust composites (Andriani et al., 2020). Second, pure BC lacks some characteristics that restrict its use in various industries. For instance, pure BC lacks magnetic, conductive, and antimicrobial

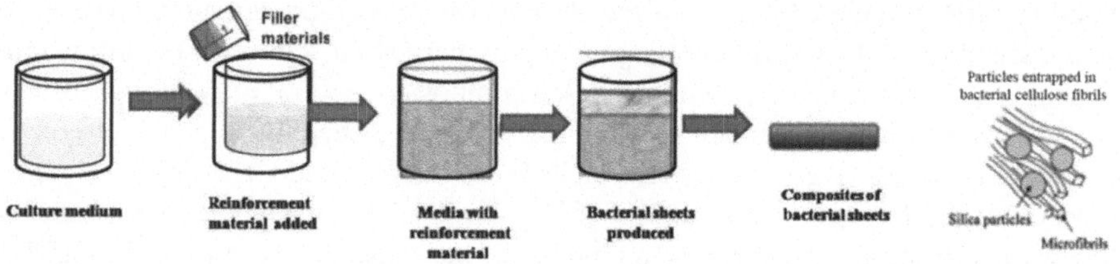

FIGURE 8.2 In situ strategy for preparation of bacterial composites.

characteristics. This is a disadvantage when designing medicinal or conductive devices for BC. Numerous findings on the preparation of BC composites with different components have been published thus far to produce BC with improved mechanical strength, biocompatibility, bactericidal activity, conductivity, and magnetic characteristics. Different techniques have been developed to synthesize BC composites depending on the intended purpose and compatibility of material impregnation.

The addition of reinforcing materials to the culture medium is the approach employed most frequently for the creation of biopolymer composites (in situ composite synthesis). Doing this with the in situ composite synthetic method, BC composites have been created using polymers such chitosan, gelatin, poly-3-hydroxybutyrate, nanomaterials, CNTs, and other solid particles such as clay and silica. The reinforcement components penetrate into the BC fibrils as they grow effectively. The antibacterial properties of reinforcing agents, which kill the BC-synthesizing cells, in addition to insolubility of certain components in culture media, have both reduced the efficacy of in situ composite synthetic technique (Figure 8.2).

The second strategy was to post-synthesize BC with reinforcing agents to address the shortcomings of the in situ process. The method is referred to as the ex situ composite synthetic approach. As was already established, the porous geometry of the BC web shape allows for the impregnation of various components within its surface. Once treated with BC films, suspensions of nanoparticles and polymer solutions can permeate the BC matrix. The compounds are chemically bonded to the cellulose chains or are physically encased inside the BC fibers, based on its chemical makeup. Ex situ synthesis has been used to make a variety of BC composites. This involves BC composites having CNTs, gelatin, polyethylene glycol, gelatin, and silver, gold, and/or chitosan nanoparticles. The primary drawback of this approach is that only particles of a certain size can enter the BC fibers. Additionally, this method cannot be used to mix BC with hydrophobic materials.

8.4 CLASSES OF BACTERIAL CELLULOSE COMPOSITES

Components varying from bulk polymers to nanopolymers were used to create BC composites. BC composites can be divided into inorganic and organic components depending on the type of reinforcement material used. These two basic types can also be broken down into BC composites having polymers, nanoparticles, metals, metal oxides, clays, and macro-sized solid particles as subclassifications. The majority of organic polymers have distinct hydrogen bonding regions. The manufacture of BC composites with many polymers of various natures is made easier by the identical nature and presence of the binding sites. Due to their unique features, metals and metal oxides constitute a distinct class of materials that have drawn a lot of interest. The exact properties required by the polymeric materials are controlled by their small size, charged nature, and individual behavior. As a result, adding metals and metal oxides to polymeric components has long been a common procedure. Metallic particles stand out for their conductivity, magnetism, photooxidizing ability, and antibacterial characteristics.

8.5 *IN SITU* SYNTHESIS OF BACTERIAL CELLULOSE COMPOSITES

This chapter is focused mainly on the in situ synthesis of BC composite. During the BC production process, various reinforcing materials such nanoparticles, chitosan, clay, glyoxal, collagen, silica nanoparticles, hydroxyapatite, poly (vinyl) alcohol, and calcium carbonate and gelatin are incorporated into the culture medium using the in situ approach. Ultimately, the BC matrix incorporates these reinforcement materials. BC composites are produced as a result of the encaged components joining the BC fibril network. The antibacterial qualities of reinforcement material have the ability to kill the BC producing cells and result in the insolubility of diverse components in culture media, have impeded the in situ process. In this method, reinforcing chemicals are added to the culture medium before the operation even starts. BC microfibrils become denser over time, forming a web-like structure. During BC synthesis, a variety of components can be imprisoned in this unusual web-shaped architecture (Eslahi et al., 2020).

Water-soluble chemicals can be added to the BC culture medium to ensure the homogeneity of composites made in situ (Mbituyimana et al., 2021). To create BC-aloe vera composite films, Saibuatong and Phisalaphong added varied amounts of aloe vera to BC synthetic media that had been cultivated under static circumstances. The BC fibrils were used to entrap the aloe vera components, and the resulting composite possessed superior physicomechanical qualities (Saibuatong and Phisalaphong, 2010). Carbon nanotubes with multiple walls were similarly put in synthetic media and cultured for two weeks. On structural examination, the CNTs were found to be trapped between the BC fibrils. By cultivating *G. xylinus* in CNT-containing culture medium and in situ hybridizing CNTs with BC, Park et al. (2015) biosynthesized a CNT-based 3D scaffold. The synthetic scaffolds were successfully implanted into the mouse calvaria to promote bone repair. In situ polyvinyl alcohol (PVA) composites were created by Gea et al. (2010) by putting a PVA solution directly into the growth medium. The optical transparency of the resulting composite is enhanced by the addition of PVA to the BC.

Using in situ composite synthesis, a BC composite with a water-insoluble poly-3-hydroxybutyrate (PBH) was created. Using a composite of BC with silver nanowires (AgNWs), Wan et al. (2009) demonstrated improved Ag^+ release, as well as a high capacity to enhance cell proliferation, skin regeneration, and epithelial tissue development. By spreading AgNWs in continuous BC, a unique step-by-step in situ biosynthesis is used to create the composite. Huang et al. used in situ fermentation, where gold nanoparticles (GNPs) with high osteogenic differentiation capacity were integrated into the BC hydrogel network (Au/BC hydrogels). Comprehensive research was done on the impacts of GNPs on the physicochemical characteristics of BC hydrogel, in vitro osteogenic differentiation, as well as in vivo bone regeneration of Au/BC hydrogels. The outcomes demonstrated that the Au/BC hydrogels could be significantly improved with better mechanical characteristics, higher porosity, greater surface area, and biocompatibility by increasing the feeding amounts of GNPs (Huang et al., 2023). The hydrogels now have exceptional biological activity in promoting osteogenic differentiation of human bone marrow-derived mesenchymal stem cells (hBMSCs). Nanobioactive glass (NBG) was used by Abdelraof et al. (2019) to create BC/NBG bioactive nanocomposites, which enhanced the production of BC. According to their research, BC/NBG composite has a wide range of applications in the biomedical field, including human bone regeneration and wound healing (Abdelraof et al., 2019; Huang et al., 2023).

Bacterial cellulose composite was also produced by incorporating carbon nanomaterial graphene oxide (GO) forming a two-dimensional nanomaterial. On its surface, it has many functional groups, that is, oxygen which links strongly to the BC, forming hydrogen bonds. The BC/GO nanocomposites with different conformations were developed that was used for absorbents for the elimination of pollutants or in energy storage devices. *G. xylinus* was used as preparation for BC/chitosan in situ composite of chitosan (poly(1,4-b-D-glucopyranosamine)) and cellulose has been explored to develop composite membranes and fibers that show various beneficial properties in biomedical areas, such as anti-inflammation, antibiosis, deodorization, drug delivery vehicles, gene delivery methods, and as orthopedic materials in wound dressing and scaffolds.

The linear sulfated polysaccharides named carrageenans were obtained from red seaweed. Bacterial cellulose (BC)/κ-carrageenan composite pellicles were prepared by using *Gluconacetobacter xylinus* ATCC53582. The gelling properties of the carrageenans were studied extensively and it was found that specific cations, such as K^+, Rb^+, and Cs^+ bind to the κ-carrageenan double helix, increasing the stability of the helix, promoting subsequent helix aggregation, and increasing gelation ability. BC/κ-carrageenan composite pellicles are used widely as additives in the food and cosmetics industries (Numata et al., 2021).

For the manufacture of BC composites for use in the biomedical field, such as for tissue engineering, collagen is used as reinforcing agent. Utilizing improved fermentation techniques and adding xanthan gum (XG) for in situ modification during submerged fermentation of *Enterobacter* sp., modified BC/XG nanocomposites with improved productivity and characteristics were produced in *Enterobacter* sp. FY-07 (FY-07) (Gao et al., 2019). The BC/XG nanocomposite outperformed BC in terms of coarser fibers, hardness, chewiness, resilience, and tensile strength, making it a better choice for food as well as other purposes. By adding 2% (w/v) alginate during the static culture of *Komagataeibacter xylinus fermentans* B-11267 for 5 days at 28°C, Liyaskina et al. formed a BC/alginate nanocomposite; they derived BC with antibiotic activity against *Staphylococcus aureus* and discovered it was suited for use as a wound treatment dressing (Liyaskina et al., 2018).

Bacterial cellulose-ciprofloxacin composites were made using the in situ inclusion approach by adding ciprofloxacin to the Hestrin-Schramm medium at the start of *Acetobacter xylinum* 0416's exponential phase of production. The static fermentation was performed for 10 days at 28 degrees Celsius. Bacteria that cause diabetic foot ulcers, such as *E. coli*, *K. pneumoniae*, and *P. aeruginosa*, were suppressed by the BC-ciprofloxacin film. A wound dressing and an antibiotic called ciprofloxacin could both be delivered *via* BC. To create a BC complex with antibacterial capabilities, Chen and colleagues undertook a series of experiments involving hydrolyzing mulberry leaves and investigating the processes of fermentation and purification. *Escherichia coli* and *Staphylococcus aureus* were used to test the antibacterial abilities of MH-BC (fermented by the mulberry leaves acid hydrolysate fermentation medium), and the experimental findings revealed also that MH-BC possesses outstanding antibacterial activities. Plant flavonoids provide its antibacterial properties, and numerous studies have been done on how it works (Chen et al., 2019).

In situ bacterial cellulose (BC)/GEL composites were created by incorporating gelatin into the culture medium that produces BC. The BC yield as well as growth rate were significantly different from those of pure BC because of the inclusion of gelatin, which conflicted with the creation of the BC pellicle form. By raising the bacterium *G. xylinum*, pure BC films were created (Chen et al., 2019). Alkali lignin (AL) is a useful component of in situ BC production that can be used in a novel way to create multifunctional composites by simultaneously promoting microbial growth and strengthening filler. In BC pores, AL self-assembles to generate primary and secondary structures that simultaneously increase toughness and thermal stability. The BC/AL films exhibit robust ultraviolet-blocking capabilities, sustained radical scavenging activity, and the ability to prevent apples from browning when they are freshly sliced, making them acceptable for food packing. An in situ technique using *K. xylinus* and functionalized graphene under static aerobic circumstances was adopted for the production of BC/reduced graphene oxide nanocomposites and increased the dispersal of 2D grapheme nano sheets in nanofiber interlayers (Figure 8.3). The percolation models used to validate the creation of an interconnected 3D network in the nanocomposites revealed an improvement in both their electrical and mechanical properties (Dhar et al., 2021).

Naeem et al. (2020) proposed the in situ self-assembly method for coating BC on the surface of banana peel fiber. *Gluconacetobacter xylinus* was fermented in a shaking environment to produce BC. Natural fibers have some limitations in the production of composite materials, mostly because of their dimensional instability, water sensitivity, and poor compatibility with many hydrophobic polymeric matrices.

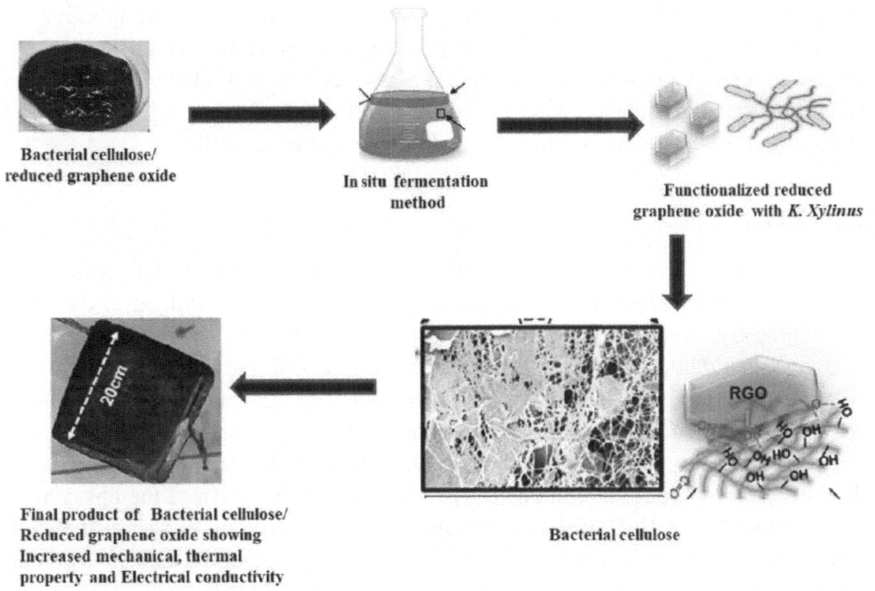

FIGURE 8.3 In situ preparation of bacterial cellulose/reduced graphene oxide nanocomposite.

Gluconacetobacter xylinus produces BC membrane while the cells are suspended in the medium. Additionally, an oxidation process was carried out by Elayaraja et al. (2017) to activate the carboxyl group of TEMPO (2,2,6,6-tetramethylpiperidine-1-oxyl radical). The TEMPO-oxidized BC was then submerged in $AgNO_3$ solution to produce AgNP-bound BC to enhance its vibriocidal action against *Vibrio parahaemolyticus* and *V. harveyi*, managing the pathogen that affects shrimp.

Bacterial cellulose and polycaprolactone (PCL) powder were added to the BC growth medium in various proportions, and the resulting mixes were then hot pressed to create a variety of BC/PCL nanocomposite films. During manufacturing, PCL powder was completely integrated into the BC network without altering its shape. The collected films revealed that PCL was evenly distributed throughout the BC. Two types of additives or reinforcing materials are utilized for in situ BC modification: interacting and noninteracting. In most of the research that has been published so far consider additives to have little impact on fermentation because of their inert nature. The traditional BC fermentation process produces gluconic acid as a byproduct when glucose is employed as the carbon source. Over time, this lowers the pH of the culture and reduces the rate of metabolism of the the microorganisms needed to create cellulose. It is crucial to keep the pH of the culture between 4.0 and 6.0, depending on the fermentation bacteria employed to produce BC (Dhar et al., 2021).

A variety of BC-based composites have been synthesized via the in situ method to incorporate functional groups and to increase the activity in different applications, including electrically conductive water purification membranes, adsorbents, free-standing electrodes, supercapacitors, tissue engineering, antimicrobials, luminescent substances, and others (Mbituyimana et al., 2021). Because of the encaged materials, the in situ approach has the advantage of improving the physicomechanical properties of BC fibrils. This method makes it very easy to create a BC-based composite with a variety of reinforcements; however, it is challenging to create an antibacterial BC composite (Saibuatong and Phisalaphong, 2010). Because the particles remain suspended in the BC synthetic fluid only briefly, composite synthesis using static BC culture is particularly challenging. At the medium-air interface in static culture, a BC sheet develops on the medium's surface. The particles cannot be captured in the BC fibers on the surface after precipitation or after migrating into the fluid. The additional components have a considerably better chance of becoming stuck in the BC fibrils in an agitated culture. Because the medium is constantly moving, the particles are prevented from settling. The most popular process uses a variety of improvements to agitating equipment and operation procedures, called in situ BC composite synthesis. However, this approach has some drawbacks that make it impossible to synthesize most BC composites. Due to their toxic effects on microbes, a number of significant bioactive (antibacterial) agents, such as Ag, ZnO, and TiO2, cannot be put directly into the medium (Shah et al., 2013; Foresti et al., 2017).

8.6 CONCLUSIONS

Bacterial cellulose is a superior biopolymer for making composite materials with a wide range of uses. Depending on the desired use, different methods have been utilized to create BC-based composites. Currently, the in situ method is most widely employed in industrialized processes because it is a simple method and a wide variety of solid and liquid reinforcement materials can be used. Fibrous materials, metallic oxides, solid clay particles, and carbonates are some examples of the various sorts of reinforcing components that have strengthened polymers while contributing a variety of biological qualities to polymeric composites. Different approaches have led to the development of novel methods for composite synthesis that have improved the physical, chemical, and biological properties of BC, ultimately enhancing their application in various fields, including biomedical, food, cosmetics, electronics, tissue engineering, gene therapy, and environmental solutions. Table 8.1 shows some applications of BC composites in various fields.

TABLE 8.1
***In Situ* Bacterial Cellulose Composites and Their Applications**

Reinforcement/Additive Material	Bacteria	Improved Properties/ Features	Applications
Collagen	*Acetobacter. xylinum*	Tissue engineering scaffold	Tissue engineering
Aloe vera gel	*Acetobacter. xylinum*	Mechanical, physical	Biomedical, wound dressing/artificial skin
Sand dollar skeleton	*Gluconacetobacter hansenii*	Porosity, density, thickness	Bone regeneration
Hydroxyapatite	*Gluconacetobacter* xylinus	Thermal property, fiber diameter and orientation, porosity	Bone regeneration
Deacetylated chitinnanocrystals	*Acetobacteraceti*	Mechanical property and antimicrobial activity	Wound dressing/artificial skin
Chitosan and heparin	*Acetobacter*xylinum	Porosity, roughness, surface area, and biocompatibility	Vascular grafts
κ-carrageenan	*Gluconacetobacter*xylinus		Food additive and cosmetics industry
Silica	-	Mechanical properties	Industrial
Inorganic particles	-	Mechanical properties	Industrial
Reduced graphene oxide	-	Conductivity	Flexible sensor
Silver nanoparticles		Antibacterial, biocide	Wound dressing, membrane
Graphene oxide	-	Antibacterial	Biomedical
Carbon nanotube	-	Mechanical strength, electrical and thermal conductivity	Electronic devices, biosensors, hydrogen storage
Gold nanoparticles	-	Conductivity	Bioelectroanalysis
Magnetic nanoparticles	-	Magnetically responsive	Microwave absorption devices and biosensors

8.7 REFERENCES

Abdelraof, M., Hasanin, M.S., Farag, M.M., and Ahmed, H.Y. (2019) Green synthesis of bacterial cellulose/bioactive glass nanocomposites: Effect of glass nanoparticles on cellulose yield, biocompatibility and antimicrobial activity. Int J Biol Macromol, 138, 975–985.

Andriani, D., Apriyana, A.Y., and Karina, M. (2020) The optimization of bacterial cellulose production and its applications: A review. Cellulose, 27(12), 6747–6766.

Chen, J., Chen, C., Liang, G., Xu, X., Hao, Q., and Sun, D. (2019) In situ preparation of bacterial cellulose with antimicrobial properties from bioconversion of mulberry leaves. Carbohydrate Polymers, 220, 170–175.

Dhar, P., Sugimura, K., Yoshioka, M., Yoshinaga, A., and Kamitakahara, H. (2021) Synthesis-property-performance relationships of multifunctional bacterial cellulose composites fermented in situ alkali lignin medium. Carbohydrate Polymers, 252, 117114.

Długa, A., Kowalonek, J., and Kaczmarek, H. (2021) Bionanocellulose/poly (Vinyl alcohol) composites produced by in-situ method and ex-situ/impregnation or sterilization methods. Materials, 14(21), 6340.

Elayaraja, S., Zagorsek, K., Li, F., and Xiang, J. (2017) In situ synthesis of silver nanoparticles into TEMPO-mediated oxidized bacterial cellulose and their antivibriocidal activity against shrimp pathogens. Carbohydrate Polymers, 166, 329–337.

Eslahi, N., Mahmoodi, A., Mahmoudi, N., Zandi, N., and Simchi, A. (2020) Processing and properties of nanofibrous bacterial cellulose-containing polymer composites: A review of recent advances for biomedical applications. Polymer Reviews, 60(1), 144–170.

Fernandes, M., and Gama, M. (2019) Development of novel bacterial cellulose composites for the textile and shoe industry. Microbial Biotechnology, 12(4), 650–661.

Foresti, M. L., Vázquez, A., and Boury, B. (2017) Applications of bacterial cellulose as precursor of carbon and composites with metal oxide, metal sulfide and metal nanoparticles: A review of recent advances. Carbohydrate Polymers, 157, 447–467.

Gabryś, T., Fryczkowska, B., Fabia, J., and Biniaś, D. (2022) Preparation of an active dressing by in situ biosynthesis of a bacterial cellulose: Graphene oxide composite. Polymers, 14(14), 2864.

Gao, M., Li, J., Bao, Z., Hu, M., Nian, R., Feng, D., and Zhang, H. (2019) A natural in situ fabrication method of functional bacterial cellulose using a microorganism. Nature Communications, 10(1), 1–10.

Gea, S., Bilotti, E., Reynolds, C.T., Soykeabkeaw, N., and Peijs, T. (2010) Bacterial cellulose-poly (vinyl alcohol) nanocomposites prepared by an in-situ process. Materials Letters, 64(8), 901–904.

Huang, C., Ye, Q., Dong, J., Li, L., Wang, M., Zhang, Y., and Jiang, Q. (2023) Biofabrication of natural Au/bacterial cellulose hydrogel for bone tissue regeneration via in-situ fermentation. Smart Materials in Medicine, 4, 1–14.

Liu, W., Du, H., Zhang, M., Liu, K., Liu, H., Xie, H., Zhang, X., and Si, C. (2020) Bacterial cellulose-based composite scaffolds for biomedical applications: A review. ACS Sustainable Chemistry & Engineering, 8(20), 7536–7562.

Liyaskina, E.V., Revin, V.V., Paramonova, E.N., Revina, N.V., and Kolesnikova, S.G. (2018) Bacterial cellulose/alginate nanocomposite for antimicrobial wound dressing. KnE Energy, 202–211.

Mbituyimana, B., Liu, L., Ye, W., Boni, B.O., Zhang, K., Chen, J., Thomas, S., Vasilievich, R.V., Shi, Z., and Yang, G. (2021) Bacterial cellulose-based composites for biomedical and cosmetic applications: Research progress and existing products. Carbohydrate Polymers, 273, 118565.

Naeem, M.A., Siddiqui, Q., Mushtaq, M., Farooq, A., Pang, Z., and Wei, Q. (2020) Insitu self-assembly of bacterial cellulose on banana fibers extracted from peels. Journal of Natural Fibers, 17(9), 1317–1328.

Numata, Y., Yoshihara, S., and Kono, H. (2021) In situ formation and post-formation treatment of bacterial cellulose/κ-carrageenan composite pellicles. Carbohydrate Polymer Technologies and Applications, 2, 100059.

Park, S., Park, J., Jo, I., Cho, S.P., Sung, D., Ryu, S., and Kim, B.S. (2015) In situ hybridization of carbon nanotubes with bacterial cellulose for three-dimensional hybrid bioscaffolds. Biomaterials, 58, 93–102.

Rajwade, J.M., Paknikar, K.M., and Kumbhar, J.V. (2015) Applications of bacterial cellulose and its composites in biomedicine. Applied Microbiology and Biotechnology, 99(6), 2491–2511.

Roy, R., Roy, R.A., and Roy, D.M. (1986) Alternative perspectives on "quasi-crystallinity": Non-uniformity and nanocomposites. Materials Letters, 4(8–9), 323–328.

Saibuatong, O.A., and Phisalaphong, M. (2010) Novo aloevera-bacterial cellulosecomposite film from biosynthesis. Carbohydrate Polymers, 79(2), 455–460.

Shah, N., Ul-Islam, M., Khattak, W.A., and Park, J.K. (2013) Overview of bacterial cellulose composites: A multipurpose advanced material. Carbohydr. Polym., 98, 1585–1598.

Singhania, R.R., Patel, A.K., Tsai, M.L., Chen, C.W., and Dong, C.D. (2021) Genetic modification for enhancing bacterial cellulose production and its applications. Bioengineered, 12(1), 6793–6807.

Tseng, Y.S., Patel, A.K., Chen, C.W., Singhania, R.R., and Dong, C.D. (2022) Improved bacterial cellulose production by *Komagataeibacter europaeus* by media engineering and optimizing culture conditions. J. Food Sci. Technol. https://doi.org/10.1007/s13197-022-05451-y

Ul-Islam, M., Khan, S., Ullah, M.W., and Park, J.K. (2015) Bacterial cellulose composites: Synthetic strategies and multiple applications in bio-medical and electro-conductive fields. Biotechnol. J., 10, 1847–1861.

Volova, T.G., Prudnikova, S.V., Kiselev, E.G., Nemtsev, I.V., Vasiliev, A.D., Kuzmin, A.P., and Shishatskaya, E.I. (2022) Bacterial Cellulose (BC) and BC composites: Production and properties. Nanomaterials, 12(2), 192.

Wan, Y.Z., Luo, H., He, F., Liang, H., Huang, Y., and Li, X.L. (2009) Mechanical, moisture absorption, and biodegradation behaviours of bacterial cellulose fibre-reinforced starch biocomposites. Comp Sci Technol, 69, 1212–1217.

Wu, Z.Y., Liang, H.W., Chen, L.F., Hu, B.C., and Yu, S.H. (2016) Bacterial cellulose: A robust platform for design of three dimensional carbon-based functional nanomaterials. Acc Chem Res, 49, 96–105.

9 Application of Bacterial Cellulose as Polymer Matrices

Olga Psaki, Mirva Sarafidou, and Apostolis Koutinas
Agricultural University of Athens
Athens, Greece

9.1 INTRODUCTION

Bacterial cellulose (BC) is a natural polymer synthesized by bacteria. It is a promising material with tremendous potential in various fields due to its properties (Hsieh et al., 2008; Yano et al., 2005). Increased interest in the use of natural polymers and given the fact that BC exhibits greater properties and potential compared with plant cellulose, there are a wide range of applications for BC. Specifically, unlike plant-derived cellulose, BC is produced without lignin and hemicellulose, resulting in high purity and crystallinity. Moreover, structurally BC is formed as a pellicle, with intra- and intermolecular hydrogen bonding and van der Waals interactions (Koizumi et al., 2008). The presence of hydroxyl groups accounts for its hydrophilic properties and high mechanical properties (Azeredo et al., 2019; Gelin et al., 2007).

New nanocomposites based on BC with desired properties have been developed by incorporating selective suitable materials into the BC matrix. The flexibility and functionality, as well as the superior mechanical properties, have made BC useful for certain applications mainly in the medical field, wastewater treatment, paper, and audio industries. Specifically, several applications of BC in the biomedical field, such as wound dressings, materials for tissue regeneration, burn treatments, and as carriers for drug delivery, have been reported (Czaja et al., 2006; Gelin et al., 2007; Swingler et al., 2021; Wahid et al., 2021). BC can be used in its pure form or as bacterial cellulose nanocomposites (BCNs) with the BC modified to incorporate other materials. Due to the porous surface traps, different types of nanomaterial can be introduced into the BC structure. Several methods have been reported for the modification of BC to produce BCNs.

Generally, ex situ, in situ, and solution of dispersed BC are the main techniques that have been used for the modification of BC. In the in situ method, the reinforcement materials are added to the BC culture media. The nanofillers of BC have a porous structure where the reinforcement materials can interact physically and chemically resulting in BCNs. The modification of BC after cultivation with the reinforcement materials is called the ex situ technique. The impregnation with different materials and polymer solution, evolve chemical and physical interactions and engrossed inside the porous BC matrix. The interaction occurs through the hydrogen bonding between BC and the reinforcement material. Solution of dispersed BC is the formation of BCNs from homogenized solutions of dissolved BC with a variety of materials (Shah et al., 2013).

Polymers are organic macromolecules formed by many smaller molecules, called monomers. Bacterial cellulose is a linear polymer consisting of D-glucopyranose units linked together by β-(1 $\rightarrow$ 4)-glycosidic bonds. Each monomer consists of three hydroxyl groups and can form intra- and intermolecular hydrogen bonds. Many biopolymers have been used together with BCNs, resulting in improvements in many physicochemical and mechanical properties (Sanyang & Jawaid, 2019).

BCNs have been synthesized for the improvement of the existing BC properties but also the addition of new features to the polymer. The impregnation with other materials can overcome the limitations that BC has and their applications to be expanded in other fields (Kim et al., 2014; Ul-Islam et al., 2015). Depending on the applications, bioactive polymers, nanomaterials, and solid particles have been used to provide specific characteristics (Kim et al., 2011; Koizumi et al., 2008).

DOI: 10.1201/9781003355434-9

The nature of the reinforcement material used for the synthesis of composites defines where the composites are applied. For instance, the utilization of biopolymers in BC sheets increase their biocompatibility, while the impregnation of magnetic materials in the BC matrix are targeted for electronics and sensor applications (Shah et al., 2013).

This chapter provides information on BC synthesis and the materials that are used for the synthesis of BC composites for enhanced applications.

9.2 PREPARATION OF BACTERIAL CELLULOSE NANOCOMPOSITES

BC has excellent mechanical properties, a high degree of crystallinity, water holding capacity, and a fibril network consisting of ultrafine nanofibers. These excellent properties make BC a good candidate for various applications, including tissue engineering, drug delivery, food application, and incorporation with other materials to produce various composites or nanocomposites (Poddar & Dikshit, 2021).

The matrix in a composite is the material that acts as a scaffold, whereas reinforcement materials are supported by the matrix and impart physicochemical and biological properties. In that way, the properties of the final composite material have been optimized because of the synergistic reaction of the independent materials (Shah et al., 2013). Because of the enhanced properties and different applications, polymer composites have gained much attention compared with each polymeric material alone. BC is used as both a matrix and reinforcing material and has been used in the development of numerous composite materials for various applications. Different routes in the field of nanotechnology have been explored to improve its mechanical, physical, and biomedical properties. BCNs are formed using materials such as polymers (Grande et al., 2009; Hu et al., 2011; Müller et al., 2013), graphene oxide (Ccorahua et al., 2017; Ma et al., 2016), carbon nanotubes (Yoon et al., 2006), metals, and metal oxides (Gea et al., 2007). Generally, the toughness of a polymer is connected with high stress and elongation but it is not always possible to improve both parameters simultaneously. For example, the incorporation of high content nanoparticles can increase stress and modulus but decrease strain (Bai et al., 2018).

9.2.1 DIFFERENT TYPES OF BACTERIAL CELLULOSE NANOCOMPOSITES

9.2.1.1 Bacterial Cellulose/Polymers Composites

Polymers are organic in nature with long, repeating chains of molecules where BCNs can be synthesized due to their similar nature. Numerous examples of BCNs using biopolymers, with advanced properties compared with BC, have been reported in the literature.

Composites of BC and chitosan have shown improvements in physicomechanical and biomedical properties. Chitosan possesses antimicrobial, antiviral, and wound-healing properties that can overcome the limitation of the BC applications in many fields (Kim et al., 2011; Lin et al., 2013; Ul-Islam et al., 2011). BC and chitosan composite can be synthesized by both in situ and ex situ methods. The improvement of mechanical properties, water holding capacity, and release rate were reported with 1% chitosan dissolved in a 1% acetic solution and employed at BC sheets, where can be further use in biomedical field (Ul-Islam et al., 2011).

Other polymers such as gelatin have been used for BCNs. The BC gelatin nanocomposites are mainly used in medicine because they improve cell adhesion.

Collagen is also a polymer that exists in nature in fibers formed in various animal tissues and organs. A BC-collagen composite resulted in better thermal stability and an increment in the elastic modulus and tensile strength, making it suitable for wound dressings (Zhijiang & Guang, 2011). Luo et al. (2008) reported that the addition of collagen in BC by using the in situs method led to decrease in the crystallinity, porosity, and thickness of the BC.

Conducting polymers are organic materials that can conduct electricity due to their unique structure where the π-bonding is highly delocalized and polarized and the biggest advantage is that they can be easily treated. The most known conducting materials are polyacetylene (PA), polyaniline (PAni), polypyrrole (PYY), poly-thiophene (PTH), poly(para-phenylene) (PPP), poly(-phenylenevinylene) (PPV), and polyfuran (PF) (Namsheer & Rout, 2021). The application of these polymers into BC creates flexible nanocomposites. Research has shown that the polymerization of PAni in a solution of BC pellicles in an in situ process resulted in high conductivity (Hu et al., 2011).

9.2.1.2 Bacterial Cellulose/Graphene Derivative Composites

Graphene can be produced from carbon by different techniques (top down and bottom up techniques) and presents outstanding mechanical properties, large specific surface area, and electrical and thermal conductivity (Sahu et al., 2021), while graphene oxide shows a high hydrophilic behavior that promotes stable aqueous colloids and the buildup of macroscopic structures (Pei & Cheng, 2012). Another graphene derivative is carbon nanotubes, which are formed in nanotubes fulfilled by graphene sheets. The combination of carbon nanotubes and BC leads to enhancement of the mechanical strength, and electrical and thermal conductivity (Torres et al., 2019).

Generally, the addition of graphene and its derivatives in BC nanocomposites enhanced multifunctional properties, resulting in their application in many fields. BC/graphene composite films were prepared by homogenization and ultrasonication and their properties were investigated (Shao et al., 2016). The mechanical properties were improved compared to the pure BC, while the water retention capacity was decreased. Graphene nanocomposite hydrogels were obtained by the addition of graphene in BC medium using a layer-by-layer interface culture strategy. The graphene/BC nanocomposite hydrogels show improvement in tensile and modulus strength of more than 91 and 279%, respectively (Luo et al., 2018). A vacuum-assisted self-assembly technique was applied for the formation of BC and graphene oxide films, resulting in the increase in Young's modulus and tensile strength of the composite film by 10% and 20%, respectively (Feng et al., 2012).

9.2.1.3 BC/Metal and Metal Oxide Composites

Metals and metal oxides are a particular class of materials that have great importance due to their excellent properties and have been studied for over 50 years. These (metal and metal oxides) are charged small-sized material whose individual behavior determines the specific properties required for polymeric materials (Shah et al., 2013). More precisely, metals and metal oxides are incorporated into polymeric materials, by the addition of metallic particles to their conductive and magnetic capabilities, photooxidative activities, and antimicrobial properties (Qian & Hinestroza, 2004). Metals and metal oxides have been used extensively in polymer composites covering a wide variety of advanced research, to introduce electrical, optical, conductive, and antimicrobial properties into polymers in order to improve the physical and mechanical properties of polymers (Bloor et al., 2005; Qian & Hinestroza, 2004).

Bacterial cellulose can also be employed to produce composites with the addition of metals and metal oxides using a wide variety of innovative synthetic approaches (Shah et al., 2013).

BC/Ag nanocomposites were synthesized by immersing BC into silver nitrate solution (Maneerung et al., 2008). The absorbed silver ion reduced in BC to silver nanoparticles by sodium borohydride. The composites produced showed high antibacterial effect. Similar research had the same observation about the high antibacterial effects in BC/Ag nanocomposites (Maria et al., 2010; Sureshkumar et al., 2010).

Several studies have been reported for BC/hydroxyapatite (Hap) composites produced through different techniques. Hap is created by calcium phosphate and has similar properties to bone minerals, leading to its use in bone grafting and drug delivery (Busuioc et al., 2016). BC/Hap nanocomposites are suitable for biomedical applications because of their resemble to human trabecular and cancellous bones (Hutchens et al., 2006; Saska et al., 2011; Zimmermann et al., 2011).

Also, composites of BC with metallic oxides were prepared for the enhancement of biological and conducting properties. For example, BC/titanium dioxide nanocomposite was prepared by the sol gel method in which, due to the strong hydrogen bonding interaction, the titanium dioxide held to the BC fibrils. The conductive behavior of the nanocomposite was examined by applying different voltages, and the results indicated that the BC/titanium dioxide surface responded to applied voltage (Gutierrez et al., 2012).

Additionally, among the metals and metallic oxides, solid materials have been used due to their biocompatibility, nontoxic nature, and mechanical reinforcing capability. Solid materials are enriched in minerals and are mainly micro to nanoparticles (Ul-Islam et al., 2012). The potential applications in various fields of polymer nanocomposites with solid materials have been thoroughly examined. Nanoparticles such as silica (Si), montmorillonite, and C_aCO_3 have been mentioned to be incorporated into the BC matrix. Montmorillonite (MMT) belongs to the smectite family and is the most used nanoparticle due to its ability to improve up to 10-fold the mechanical properties of the composite (Liu et al., 2011). The composites of MMT and others clays can be applied in biomedical fields because they do not require the use of toxic or expensive organic solvents (Giannelis et al., 1999). Also, they have a wide range of application in the medical field, where effects such as skin protection, cleansing, and antibacterial activity have been shown (Haydel et al., 2008). Another solid material, silica, has been used in the preparation of BC composites. The presence of the silica particles in BC composite has a different effect on the structure and mechanical properties. The differences between the composites are owing to the impregnation strategies used (Ashori et al., 2012; Yano et al., 2008).

9.2.2 Methods for the Preparation of Bacterial Cellulose Nanocomposites

The preparation of BCNs involves the production of BC and the incorporation of nanofillers for multifunctional properties. These steps can be performed simultaneously (one-step preparation) or separately in different steps (two-step preparation) using dispersed BC or macroscopic BC gel. Due to the large number of hydroxyl groups on the surface of BC, which leads to inherent hydrophilicity, an intermediate step may be required to convert it into a hydrophobic material. In this way, the presence of a large number of hydroxyl groups enables different types of surface modifications that extend its use for other applications (Sun et al., 2022; Torres et al., 2019).

9.2.2.1 In Situ and ex Situ Production

The in situ method involves the addition of the reinforcing materials to the culture medium at the beginning of the BC synthesis process. BC microfibrils become denser over time and form a reticular structure, which can include various materials added to the synthetic BC medium (Ul-Islam et al., 2012). The additives become part of the BC fibril network, resulting in BCNs. Figure 9.1a shows an example of in situ BCN production in static culture, but they can be prepared also in agitated culture. In static culture, the particles do not remain suspended in the synthetic BC media for long, so they cannot be entrapped in the BC fibers at the surface. Stirred culture, however, provides a much better environment for the added substances to become entrapped in the BC fibrils, as the particles do not settle due to the constant agitation of the medium. In summary, in situ production of BCNs is the most widely used approach, employing a variety of modifications to stirring devices and working methods (Hu et al., 2014).

The ex situ method involves modifications after the BC produced in the fermentation medium has been purified (Shah et al., 2013). The BC is impregnated with other materials as additives to produce BCNs (Figure 9.1b). Liquid materials and tiny solid particles can easily penetrate and settle into the porous BC matrix. The properties of BC are affected differently by the additives used. The major hurdle in the ex situ method is the size and nature of the additives. Only submicron to nanoscale materials can be impregnated into the BC matrix, because larger particles cannot penetrate the BC pores and hydrophobic materials cannot bind with BC. In addition, the structural arrangement of BC fibrils is not always uniform. Moreover, the penetrating materials may not be uniformly distributed in the BC matrix. Therefore, new synthetic routes for BC composites need to be found to solve this problem (Sun et al., 2022; Fernandes et al., 2020).

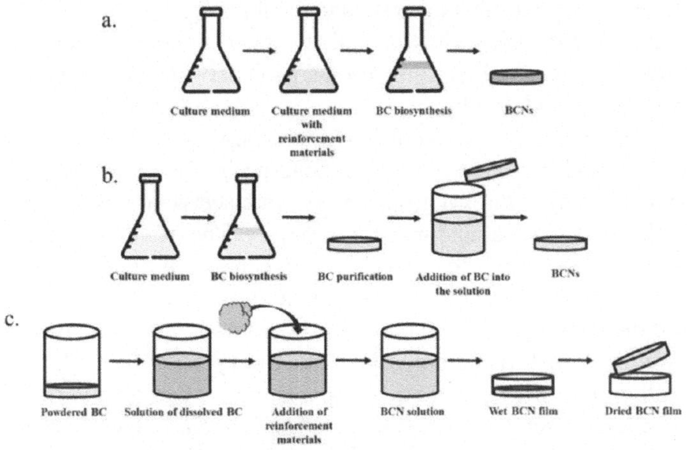

FIGURE 9.1 (a) Schematic representation of bacterial cellulose nanocomposites (BCNs) produced by the in situ method. Particles trapped between growing bacterial cellulose (BC) fibrils are visible in the figure. (b) Schematic representation of BCNs prepared by the ex situ method. The example illustrates the penetration of particles into the BC matrix. (c) Schematic representation of BCNs prepared from dissolved BC solutions. The composite solution was cast to produce BC films.

9.2.2.2 Solution of Dispersed Bacterial Cellulose

Bacterial cellulose presents a three-dimensional ultrafine network with average diameter values lower than 100 nm and length of around 100 μm (Choi & Shin, 2020). With the use of appropriate chemical, mechanical, and/or enzymatic methods, cellulose can be further reduced in size in diameter or both diameter and length, up to the nanoscale (Mondal, 2017). Depending on the dimension of the BC fibrils, the particles can be characterized as cellulose microfibrils (CMF), or microcrystalline cellulose (MCC) and cellulose nanofibrils (CNF), nano-fibrillated cellulose, or nanocrystalline cellulose (NCC). CMFs are cellulose structures of several micrometers in length and dimension, whereas CNF have at least one dimension in the nanometer range. The dispersed BCNs with a high length-to-diameter ratio could reinforce the polymer matrix in a similar manner as inorganic particles (Pei & Cheng, 2012).

Regarding the different BC particles that can be prepared, an effective strategy to synthesize a wide range of BCNs is solution mixing. The BC particles are dissolved in a specific solvent such as ionic liquids or sodium hydroxide, and then the desired materials (e.g., polymers, nanoparticles, and carbon nanotubes) are added to the prepared solution to obtain a homogeneous blend. The mixture is then poured onto a clean glass surface to form composite materials after the solvent evaporates (Figure 9.1c). The main reason for using these solvents is the limited solubility of BC in water and most widely used organic solvents (Lindman et al., 2010). Gao et al. (2011) prepared regenerated BC films from BC dissolved in *N*-methy-morpholine-*N*-oxide (NMMO) and found that they exhibited a uniform fibril arrangement with excellent mechanical and thermal properties. The composite fabricated by this method exhibits stronger interactions between BC and the combining material (Shah et al., 2013).

9.2.2.3 Surface Modification

Because of the severe microbial fermentation environment, modification of BC by biosynthesis is not always possible. In some cases, chemically modified methods are more suitable for BC material compared to the biosynthesis method. Since the chemical composition of BC is similar to plant cellulose, the same modifications such as acetylation, carboxymethylation, phosphorylation, alkylation, sulfation, and other graft copolymerization and crosslinking reactions can be used to obtain a range of BC derivatives. By introducing new functional groups into the BC structure, BC can be provided with various features such as hydrophobicity, ion adsorption capacity, and optical properties, while maintaining the unique three-dimensional nano network and excellent mechanical properties of BC (Sanyang & Jawaid, 2019).

The modification of BC for the production of nanocomposites can be fulfilled using different methods and numerous additives or materials, depending on the required application (Hu et al., 2014). Acetylation replaces the hydroxyl groups of BC with acetyl groups, resulting in a less hydrophilic molecule with better thermal and moisture stability. The extent of acetylation can be influenced by various parameters during the process, such as reaction time, temperature, and catalyst content (Ramírez et al., 2016). Lin et al. (2014) prepared a blend of bacterial carboxymethyl cellulose (BCMC)/sodium alginate by ionic crosslinking with Ca^{2+}, which showed improvement in tensile strength, swelling, and thermal properties. In carboxymethylation, the hydroxyl groups are replaced by carboxyl groups, which promote the appearance of coordination sites. Methylcellulose (MC) is a form of cellulose that is produced by its etherification. In the synthesis of MC, cellulose and an etherification agent, such as methyl chloride, methyl iodine, or dimethyl sulfate, are reacted in an alkaline medium. In industry, MC is produced in a heterogeneous medium with methyl chloride. Polymer Recycling Group produced MC in heterogeneous phase in an alkaline medium and used dimethyl sulfate as a methylating agent (Oliveira et al., 2015).

9.2.3 Structural and Mechanical Characterizations of Bacterial Cellulose Nanocomposites

The properties of BCNs vary through the different applications, so they must be characterized in many different ways. X-ray diffraction (XRD) and Fourier transform infrared spectroscopy (FTIR) are methods used to characterize the possible interactions between components in BCNs. The percentage of crystallinity of the structure can be determined by XRD, but also the phase composition of the materials. The crystallinity patterns of both the BC and BCNs can be characterized to determine the difference between the two materials. To identify the functional groups of native BC and BCNs, FTIR is an effective analytical tool to characterize the difference between the starting and the final material. The FTIR spectra provides a profile of the sample, a distinctive molecular fingerprint that can be used to screen and scan samples for many different constituents. Tensile properties indicate how materials respond to tensile forces. Tensile testing is an important test for determining the elastic modulus, elongation, and Young's modulus of native BC and BC composites. The results of the test are used to select materials for different purposes and applications, taking into account how they respond to different types of forces. The study of the microstructure of BC composites provides information about the appearance of the manufactured cellulose at the microscopic level. A field emission scanning electron microscope (FESEM) is usually used to study the microstructure of BC. Since BC consists of multiple nanocellulose networks, high-magnification instruments such as the FESEM provide a better overview of the samples.

9.3 APPLICATIONS

BCNs are ideal building blocks due to their unique structure, which gives rise to properties such as high Young's modulus and enhanced mechanical strength. The physical and chemical properties of bacterial cellulose have enabled its use in both hydrophilic and hydrophobic composite matrices, including films, hydrogels, and aerogels. Its incorporation into final products determines the final properties according to the morphology and surface properties of the former. Ultimate biodegradable BC-based materials could be fabricated to serve sectors where there is high demand, in the form of aerogels, hydrogels, micro- and nanoparticles, nanocomposites, emulsions, and membranes/films (Thomas et al., 2018). BCNs are used in many fields, such as biomedicine, the food industry, acoustic membranes, functional paper, nanostructured biomaterials, and biocomposites (Figure 9.2).

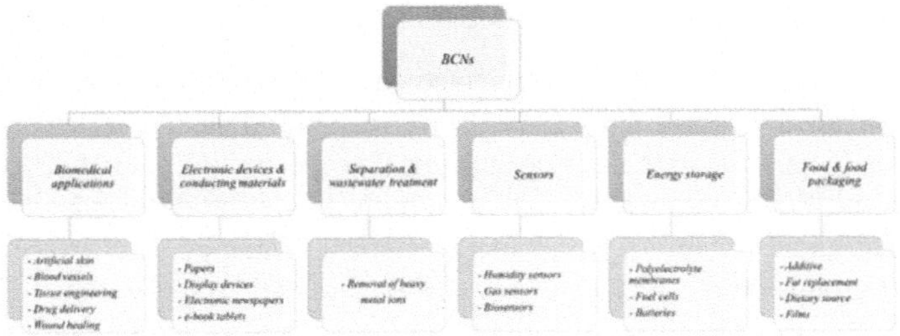

FIGURE 9.2 Illustration of applications for bacterial cellulose nanocomposites (BCNs) in different fields.

9.3.1 Biomedical Applications

As already mentioned, BCNs can be produced by adding selected bioactive material into the BC either during or after fermentation to modify the structure. The antimicrobial activity and conductive properties of BCNs make them suitable for biomedical applications. The final material could have improved properties, such as good mechanical properties and high moisture-retention properties, essential characteristics of coating materials for treating different kinds of wounds, ulcers, and burns (Legeza et al., 2004; Ul-Islam et al., 2012). BCNs have also been used for artificial skin, blood vessels, scaffolds for tissue engineering, and drug delivery (Klemm et al., 2001; Charpentier et al., 2006; Arias et al., 2016; Amin et al., 2012; Müller et al., 2013). Specifically, chitosan, silver nanoparticles, ZnO, TiO_2, and montmorillonite have been used for antimicrobial wound dressings (Khan et al., 2015; Moritz et al., 2014; Ul-Islam et al., 2014; Lin et al., 2013; Ul-Islam et al., 2012; Maneerung et al., 2008). Alginates, calcium phosphates, gelatin methacryloyl, and silk fibroin have been used as scaffolds for tissue engineering (Barud et al., 2015; Kirdponpattara et al., 2015). Also, carboxymethylcellulose and graphene oxide have been used to prepare controlled-release hydrogels (Juncu et al., 2016; Luo et al., 2017).

9.3.2 Electronic Devices and Conducting Materials

The use of BC nanocomposites has been explored as conductive materials for electrical and electronic devices such as papers and display devices (Müller et al., 2012; Lee et al., 2012). The addition of many conducting nanomaterials has been reported in previous studies (Yoon et al., 2006; Feng et al., 2012; Zhou et al., 2013). Müller et al. (2011) introduced pyrrole into the BC matrix by in situ oxidative polymerization, resulting in an electrical conductivity of 1 S cm^{-1} for the composite. The incorporation of polyaniline into a BC matrix results in improved conductive properties used in signal receivers, as well as to control cell function in tissue engineering (Paâ et al., 2013; Müller et al., 2011). Nogi et al. (2005) prepared BCNs with acrylic resins by impregnating BC sheets into an acrylic resin solution. This resulted in transparent, stronger composites and showed stable permeability with temperature changes. Since BC is not conductive by nature, it is first converted into electrically conductive films by incorporating some conductive materials. Then, they are immobilized with electrochromic dyes and attached to electrodes, where application of a potential results in a reversible color change (Shah & Brown, 2005). These display devices are very flexible, biodegradable, high contrast, and have high reflectivity. Organic light-emitting diodes (OLEDs) devices based on flexible BC membranes have already been developed (Legnani et al., 2008) and are suitable for numerous applications, including electronic newspapers, e-book tablets, rewritable maps, dynamic wall newspapers, and learning tools (Shah & Brown, 2005). Moreover, palladium, polypyrrole and molybdenum disulphide have been used for batteries applications (Evans et al., 2003; Zhang et al., 2015, 2016), titanium dioxide, graphite and polyaniline has been used for electric and electronic instruments (Gutierrez et al., 2012; Kiziltas et al., 2016; Hu et al., 2011), and gold and platinum for the production of biosensors (Li et al., 2011; Foresti et al., 2017).

9.3.3 Separation and Wastewater Treatment

Modification using acrylic acid with ion exchange capacity was used to produce BC composite membranes that had an absorption capacity for heavy metals and metal ions and an additional ability to separate trace metals due to their electrochemical properties (Choi et al., 2004). In the case of wastewater treatment, BCNs with porous structure, high tensile strength, and water holding capacity could be used as an adsorbent for heavy metal removal (Wang et al., 2015). Also, the presence of many hydroxyl groups on the surface of BC makes it effective for the removal of heavy metal ions. Nonetheless, BC, to which no functional groups have been added to modify and improve its properties, has several disadvantages, such as low adsorption capacity, poor selectivity, and high hydrophilicity, making it swell readily in water. Some examples of biosorbents prepared ex situ are polyethylenimine-coated BC, amino BC, ammonium sulfamate BC, spherical iron oxide BC, and carboxymethylated BC.

9.3.4 Sensors

The main properties of BC, such as the number of hydroxyl groups and good water absorption, make it a prime material for building high-performance humidity or gas sensors. Combined with a quartz crystal microbalance (QCM) system, the BC nanofiber membrane coating can be a good solution for constructing highly sensitive sensors. Hu et al. (2011) fabricated a highly stable and sensitive humidity sensor based on BC-coated QCM, which exhibits good sensing characteristics by increasing by more than two orders of magnitude with increasing relative humidity and good reversible behavior and long-term stability. In gas sensors, a formaldehyde sensor based on nanofibrous polyethyleneimine (PEI)/BC membranes coated with QCM was fabricated. The sensor showed high sensitivity, with good linearity, selectivity, and repeatability and low detection limit (Hu et al., 2011). The structure, stability, and biocompatibility of BC make it a good candidate for biosensor construction. Au-BC nanocomposites have been prepared for enzyme immobilization and glucose biosensor fabrication, even for monitoring in real blood samples (Zhang et al., 2010; Wang et al., 2010).

9.3.5 Energy Storage

Energy storage is another application where BCNs can be used. Structures with very large surface areas suitable for these applications can be fabricated because of the ability of BCs to catalyze palladium (Pd). Yang et al. (2009) developed a BC/Pt composite by in situ reduction that showed high electrocatalytic activity, whereas BC films doped with protonic acids exhibited high proton conductivity. Consequently, BC is suitable for the development of polyelectrolyte membranes and fuel cells.

Recently, BC has been used to improve battery performance. Zhang et al. (2022) used BC as a substrate and fabricated a flexible Ni-containing coordination polymer-modified carbonized BC (Ni-CBC) interlayer for lithium-sulfur (Li-S) batteries. This addition gives to the batteries stable discharge performance at 0.2C during long charge-discharge cycles and high capacity of 852 mAh g-1 at 2C. Ma et al. (2021) prepared a carbon nanofiber aerogel by carbonization of BC nanofibers using surface carboxylation treatment, which exhibited a specific capacitance of 268 F g-1 at 0.5 A g-1, among the best values of self-supporting carbon materials.

9.3.6 Food and Food Packaging

BC is a dietary fiber, often used in the food industry because of its rheological properties, suspending, water retention, thickening and emulsifying stability. Also, another advantage is the ability to easily modify the BC with the addition of color and flavor or change the shape and texture. BC can also be used as an additive, fat replacement, dietary source, and it is a suitable replacement for gelatin. Cellulose has been applied usually as reinforcement materials as well as matrices for the development of films for food packaging applications (Azeredo et al., 2017).

BC is widely used in the food industry because of its compatibility with several food products. Cellulose-based films and coatings are used for packaging and wrapping because it is easy to control the permeability of water vapor and CO_2 and enhance the mechanical strength of the food packaging (Tomé et al., 2010). Modification of BC film by the incorporation of sorbic acid as antibacterial agent was developed. The antimicrobial effect of the film was tested against *Escherichia coli* K12-MG165 and was shown to have promising antimicrobial properties. It was suggested that the antimicrobial activity is strongly affected by the concentration and water solubility of the sorbic acid (Wang et al., 2018). A new label sensor was also developed for chicken storage by the immobilization of red methyl in BC, using the absorption technique. Evaluation of the freshness of chicken cuts was investigated using changes in the sensor color corresponding to the length of storage. Also, the evaluation of freshness of food was tested in BC/curcumin film and BC/bromophenol blue film. The freshness of the foods was examined with the films as color indicators during the storage. A pH indicator was produced by immersing the BC into anthocyanin solution, where the films responded to pH 2 to 10 (Kuswandi et al., 2012; Pourjavaher et al., 2017). The ability to control aerobic bacteria, mainly *Listeria monocytogenes,* on the surface of vacuum-packaged frankfurters was evaluated using BC membranes impregnated with nisin. The results were a decrease of *Listeria monocytogenes* total bacteria after 14 days of storage compared to uncoated samples (Nguyen et al., 2008). Additionally, BNCs were developed with the addition of 7.5% cellulose nanocrystals and 0.6% TiO_2 in wheat gluten coatings for unbleached craft paper sheets. The antimicrobial activity of coated papers was enhanced, with a 98.5% reduction in CFU, showing that TiO_2 improved the properties. Also, an improvement in mechanical properties was observed (El-Wakil et al., 2015). The functionalization of nanocellulose with silver nanoparticles (AgNPs) as hydrogels, aerogels, and films was studied for antibacterial properties where it showed that it could be used for active food packaging applications (Dong et al., 2013).

9.4 CONCLUSIONS

Bacterial cellulose is a versatile polymer than can be applied in many fields due to its structure and properties. However, to overcome some limitations to the applications of pure BC, BC-composites have been developed. This chapter discusses BC composite preparation, using a variety of reinforcement materials by applying different techniques. Also, recent studies of the applications of BC composites in different fields have been discussed. BC turns out to be a functional tool that can be easily modified for the enhancement of its properties or to import new properties due to the reinforcement material. Meanwhile the development on a large scale is a very promising aspect due to the low environmental and health impact.

9.5 REFERENCES

Amin, M. C. I. M., Abadi, A. G., Ahmad, N., Katas, H., & Jamal, J. A. 2012. Bacterial cellulose film coating as drug delivery system: Physicochemical, thermal and drug release properties. *Sains Malaysiana, 41*(5), 561–568.

Arias, S. L., Shetty, A. R., Senpan, A., Echeverry-Rendón, M., Reece, L. M., & Allain, J. P. 2016. Fabrication of a functionalized magnetic bacterial nanocellulose with iron oxide nanoparticles. *JoVE (Journal of Visualized Experiments)*, (111), e52951.

Ashori, A., Sheykhnazari, S., Tabarsa, T., Shakeri, A., & Golalipour, M. 2012. Bacterial cellulose/silica nanocomposites: Preparation and characterization. *Carbohydrate Polymers, 90*(1), 413–418.

Azeredo, H. M., Barud, H., Farinas, C. S., Vasconcellos, V. M., & Claro, A. M. 2019. Bacterial cellulose as a raw material for food and food packaging applications. *Frontiers in Sustainable Food Systems, 3,* 7.

Azeredo, H. M., Rosa, M. F., & Mattoso, L. H. C. 2017. Nanocellulose in bio-based food packaging applications. *Industrial Crops and Products, 97,* 664–671.

Bai, C., Huang, X., Xie, F., & Xiong, X. 2018. Microcrystalline cellulose surface-modified with acrylamide for reinforcement of hydrogels. *ACS Sustainable Chemistry & Engineering*, *6*(9), 12320–12327.

Barud, H. O., Barud, H. D. S., Cavicchioli, M., do Amaral, T. S., de Oliveira, Jr., O. B., Santos, D. M., . . . Ribeiro, S. J. 2015. Preparation and characterization of a bacterial cellulose/silk fibroin sponge scaffold for tissue regeneration. *Carbohydrate Polymers*, *128*, 41–51.

Bloor, D., Donnelly, K., Hands, P. J., Laughlin, P., & Lussey, D. 2005. A metal: Polymer composite with unusual properties. *Journal of Physics D: Applied Physics*, *38*(16), 2851.

Busuioc, C., Stroescu, M., Stoica-Guzun, A., Voicu, G., & Jinga, S. I. 2016. Fabrication of 3D calcium phosphates based scaffolds using bacterial cellulose as template. *Ceramics International*, *42*(14), 15449–15458.

Ccorahua, R., Troncoso, O. P., Rodriguez, S., Lopez, D., & Torres, F. G. 2017. Hydrazine treatment improves conductivity of bacterial cellulose/graphene nanocomposites obtained by a novel processing method. *Carbohydrate Polymers*, *171*, 68–76.

Charpentier, P. A., Maguire, A., & Wan, W. K. 2006. Surface modification of polyester to produce a bacterial cellulose-based vascular prosthetic device. *Applied Surface Science*, *252*(18), 6360–6367.

Choi, S. M., & Shin, E. J. 2020. The nanofication and functionalization of bacterial cellulose and its applications. *Nanomaterials*, *10*(3), 406.

Choi, Y. J., Ahn, Y., Kang, M. S., Jun, H. K., Kim, I. S., & Moon, S. H. 2004. Preparation and characterization of acrylic acid-treated bacterial cellulose cation-exchange membrane. *Journal of Chemical Technology & Biotechnology: International Research in Process, Environmental & Clean Technology*, *79*(1), 79–84.

Czaja, W., Krystynowicz, A., Bielecki, S., & Brown, Jr., R. M. 2006. Microbial cellulose: The natural power to heal wounds. *Biomaterials*, *27*(2), 145–151.

Dong, H., Snyder, J. F., Tran, D. T., & Leadore, J. L. 2013. Hydrogel, aerogel and film of cellulose nanofibrils functionalized with silver nanoparticles. *Carbohydrate Polymers*, *95*(2), 760–767.

El-Wakil, N. A., Hassan, E. A., Abou-Zeid, R. E., & Dufresne, A. 2015. Development of wheat gluten/nanocellulose/titanium dioxide nanocomposites for active food packaging. *Carbohydrate Polymers*, *124*, 337–346.

Evans, B. R., O'Neill, H. M., Malyvanh, V. P., Lee, I., & Woodward, J. 2003. Palladium-bacterial cellulose membranes for fuel cells. *Biosensors and Bioelectronics*, *18*(7), 917–923.

Feng, Y., Zhang, X., Shen, Y., Yoshino, K., & Feng, W. 2012. A mechanically strong, flexible and conductive film based on bacterial cellulose/graphene nanocomposite. *Carbohydrate Polymers*, *87*(1), 644–649.

Fernandes, I. D. A. A., Pedro, A. C., Ribeiro, V. R., Bortolini, D. G., Ozaki, M. S. C., Maciel, G. M., & Haminiuk, C. W. I. 2020. Bacterial cellulose: From production optimization to new applications. *International Journal of Biological Macromolecules*, *164*, 2598–2611.

Foresti, M. L., Vázquez, A., & Boury, B. 2017. Applications of bacterial cellulose as precursor of carbon and composites with metal oxide, metal sulfide and metal nanoparticles: A review of recent advances. *Carbohydrate Polymers*, *157*, 447–467.

Gao, Q., Shen, X., & Lu, X. 2011. Regenerated bacterial cellulose fibers prepared by the NMMO·H2O process. *Carbohydrate Polymers*, *83*(3), 1253–1256.

Gea, S., Torres, F. G., Troncoso, O. P., Reynolds, C. T., Vilasecca, F., Iguchi, M., & Peijs, T. 2007. Biocomposites based on bacterial cellulose and apple and radish pulp. *International Polymer Processing*, *22*(5), 497–501.

Gelin, K., Bodin, A., Gatenholm, P., Mihranyan, A., Edwards, K., & Strømme, M. 2007. Characterization of water in bacterial cellulose using dielectric spectroscopy and electron microscopy. *Polymer*, *48*(26), 7623–7631.

Giannelis, E. P., Krishnamoorti, R., & Manias, E. 1999. Polymer-silicate nanocomposites: Model systems for confined polymers and polymer brushes. *Polymers in Confined Environments*, 107–147.

Grande, C. J., Torres, F. G., Gomez, C. M., Troncoso, O. P., Canet-Ferrer, J., & Martínez-Pastor, J. 2009. Development of self-assembled bacterial cellulose: Starch nanocomposites. *Materials Science and Engineering: C*, *29*(4), 1098–1104.

Gutierrez, J., Tercjak, A., Algar, I., Retegi, A., & Mondragon, I. 2012. Conductive properties of TiO2/bacterial cellulose hybrid fibres. *Journal of Colloid and Interface Science*, *377*(1), 88–93.

Haydel, S. E., Remenih, C. M., & Williams, L. B. 2008. Broad-spectrum in vitro antibacterial activities of clay minerals against antibiotic-susceptible and antibiotic-resistant bacterial pathogens. *Journal of Antimicrobial Chemotherapy*, *61*(2), 353–361.

Hsieh, Y. C., Yano, H., Nogi, M., & Eichhorn, S. J. 2008. An estimation of the Young's modulus of bacterial cellulose filaments. *Cellulose*, *15*(4), 507–513.

Hu, W., Chen, S., Yang, J., Li, Z., & Wang, H. 2014. Functionalized bacterial cellulose derivatives and nanocomposites. *Carbohydrate Polymers*, *101*, 1043–1060.

Hu, W., Chen, S., Yang, Z., Liu, L., & Wang, H. 2011. Flexible electrically conductive nanocomposite membrane based on bacterial cellulose and polyaniline. *The Journal of Physical Chemistry B*, *115*(26), 8453–8457.

Hutchens, S. A., Benson, R. S., Evans, B. R., O'Neill, H. M., & Rawn, C. J. 2006. Biomimetic synthesis of calcium-deficient hydroxyapatite in a natural hydrogel. *Biomaterials*, *27*(26), 4661–4670.

Juncu, G., Stoica-Guzun, A., Stroescu, M., Isopencu, G., & Jinga, S. I. 2016. Drug release kinetics from carboxymethylcellulose-bacterial cellulose composite films. *International Journal of Pharmaceutics*, *510*(2), 485–492.

Khan, S., Ul-Islam, M., Khattak, W. A., Ullah, M. W., & Park, J. K. 2015. Bacterial cellulose-titanium dioxide nanocomposites: Nanostructural characteristics, antibacterial mechanism, and biocompatibility. *Cellulose*, *22*(1), 565–579.

Kim, H., Nam, S., Jeong, J., Lee, S., Seo, J., Han, H., & Kim, Y. 2014. Organic solar cells based on conjugated polymers: History and recent advances. *Korean Journal of Chemical Engineering*, *31*(7), 1095–1104.

Kim, J., Cai, Z., Lee, H. S., Choi, G. S., Lee, D. H., & Jo, C. 2011. Preparation and characterization of a bacterial cellulose/chitosan composite for potential biomedical application. *Journal of Polymer Research*, *18*(4), 739–744.

Kirdponpattara, S., Khamkeaw, A., Sanchavanakit, N., Pavasant, P., & Phisalaphong, M. 2015. Structural modification and characterization of bacterial cellulose: Alginate composite scaffolds for tissue engineering. *Carbohydrate Polymers*, *132*, 146–155.

Kiziltas, E. E., Kiziltas, A., Rhodes, K., Emanetoglu, N. W., Blumentritt, M., & Gardner, D. J. 2016. Electrically conductive nano graphite-filled bacterial cellulose composites. *Carbohydrate Polymers*, *136*, 1144–1151.

Klemm, D., Schumann, D., Udhardt, U., & Marsch, S. 2001. Bacterial synthesized cellulose: Artificial blood vessels for microsurgery. *Progress in Polymer Science*, *26*(9), 1561–1603.

Koizumi, S., Yue, Z., Tomita, Y., Kondo, T., Iwase, H., Yamaguchi, D., & Hashimoto, T. 2008. Bacterium organizes hierarchical amorphous structure in microbial cellulose. *The European Physical Journal E*, *26*(1), 137–142.

Kuswandi, B., Restyana, A., Abdullah, A., Heng, L. Y., & Ahmad, M. 2012. A novel colorimetric food package label for fish spoilage based on polyaniline film. *Food Control*, *25*(1), 184–189.

Lee, B. H., Kim, H. J., & Yang, H. S. 2012. Polymerization of aniline on bacterial cellulose and characterization of bacterial cellulose/polyaniline nanocomposite films. *Current Applied Physics*, *12*(1), 75–80.

Legeza, V. I., Galenko-Yaroshevskii, V. P., Zinov'ev, E. V., Paramonov, B. A., Kreichman, G. S., Turkovskii, I. I., . . . Khripunov, A. K. 2004. Effects of new wound dressings on healing of thermal burns of the skin in acute radiation disease. *Bulletin of Experimental Biology and Medicine*, *138*(9), 311–315.

Legnani, J. C., Vilani, C., Calil, V., Barud, H., Quirino, W., Achete, C., Ribeiro, S. J. L., & Cremona, M. 2008. Bacterial cellulose membrane as flexible substrate for organic light emitting devices. *Thin Solid Films*, *517*, 1016–1020.

Li, Z., Zhu, B. J., Yang, J. X., Peng, K., Zhou, B. H., Xu, R. Q., Hu, W. L., Chen, S. Y., & Wang, H. P. 2011. Method for manufacture of bacterial cellulose hydrogel cold pack. In *Biomedical engineering: Frontiers and challenges*. Croatia. IntechOpen.

Lin, Q., Zheng, Y., Ren, L., Wu, J., Wang, H., An, J., & Fan, W. 2014. Preparation and characteristic of a sodium alginate/carboxymethylated bacterial cellulose composite with a crosslinking semi-interpenetrating network. *Journal of Applied Polymer Science*, *131*(3).

Lin, W. C., Lien, C. C., Yeh, H. J., Yu, C. M., & Hsu, S. H. 2013. Bacterial cellulose and bacterial cellulose: Chitosan membranes for wound dressing applications. *Carbohydrate Polymers*, *94*(1), 603–611.

Lindman, B., Karlström, G., & Stigsson, L. 2010. On the mechanism of dissolution of cellulose. *Journal of Molecular Liquids*, *156*(1), 76–81.

Liu, H., Chaudhary, D., Yusa, S. I., & Tadé, M. O. 2011. Glycerol/starch/Na+-montmorillonite nanocomposites: A XRD, FTIR, DSC and 1H NMR study. *Carbohydrate Polymers*, *83*(4), 1591–1597.

Luo, H., Ao, H., Li, G., Li, W., Xiong, G., Zhu, Y., & Wan, Y. 2017. Bacterial cellulose/graphene oxide nanocomposite as a novel drug delivery system. *Current Applied Physics*, *17*(2), 249–254.

Luo, H., Dong, J., Xu, X., Wang, J., Yang, Z., & Wan, Y. 2018. Exploring excellent dispersion of graphene nanosheets in three-dimensional bacterial cellulose for ultra-strong nanocomposite hydrogels. *Composites Part A: Applied Science and Manufacturing, 109*, 290–297.

Luo, H., Xiong, G., Huang, Y., He, F., Wang, Y., & Wan, Y. 2008. Preparation and characterization of a novel COL/BC composite for potential tissue engineering scaffolds. *Materials Chemistry and Physics, 110*(2–3), 193–196.

Ma, L., Liu, R., Niu, H., Zhao, M., & Huang, Y. 2016. Flexible and freestanding electrode based on polypyrrole/graphene/bacterial cellulose paper for supercapacitor. *Composites Science and Technology, 137*, 87–93.

Ma, Y., Liu, Q., Li, W., Zheng, Y., Shi, Q., Zhou, Z., . . . Fang, X. 2021. Ultralight and robust carbon nanofiber aerogels for advanced energy storage. *Journal of Materials Chemistry A, 9*(2), 900–907.

Maneerung, T., Tokura, S., & Rujiravanit, R. 2008. Impregnation of silver nanoparticles into bacterial cellulose for antimicrobial wound dressing. *Carbohydrate Polymers, 72*(1), 43–51.

Maria, L., Santos, A. L., Oliveira, P. C., Valle, A. S., Barud, H. S., Messaddeq, Y., & Ribeiro, S. J. 2010. Preparation and antibacterial activity of silver nanoparticles impregnated in bacterial cellulose. *Polimeros, 20*, 72–77.

Mondal, S. 2017. Preparation, properties and applications of nanocellulosic materials. *Carbohydrate Polymers, 163*, 301–316.

Moritz, S., Wiegand, C., Wesarg, F., Hessler, N., Müller, F. A., Kralisch, D., . . . Fischer, D. 2014. Active wound dressings based on bacterial nanocellulose as drug delivery system for octenidine. *International Journal of Pharmaceutics, 471*(1–2), 45–55.

Müller, A., Ni, Z., Hessler, N., Wesarg, F., Müller, F. A., Kralisch, D., & Fischer, D. 2013. The biopolymer bacterial nanocellulose as drug delivery system: Investigation of drug loading and release using the model protein albumin. *Journal of Pharmaceutical Sciences, 102*(2), 579–592.

Müller, D., Mandelli, J. S., Marins, J. A., Soares, B. G., Porto, L. M., Rambo, C. R., & Barra, G. M. O. 2012. Electrically conducting nanocomposites: Preparation and properties of polyaniline (PAni)-coated bacterial cellulose nanofibers (BC). *Cellulose, 19*(5), 1645–1654.

Müller, D., Rambo, C. R., Recouvreux, D. O. S., Porto, L. M., & Barra, G. M. O. 2011. Chemical in situ polymerization of polypyrrole on bacterial cellulose nanofibers. *Synthetic Metals, 161*(1–2), 106–111.

Namsheer, K., & Rout, C. S. 2021. Conducting polymers: A comprehensive review on recent advances in synthesis, properties and applications. *RSC Advances, 11*(10), 5659–5697.

Nguyen, V. T., Gidley, M. J., & Dykes, G. A. 2008. Potential of a nisin-containing bacterial cellulose film to inhibit Listeria monocytogenes on processed meats. *Food Microbiology, 25*(3), 471–478.

Nogi, M., Handa, K., Nakagaito, A. N., & Yano, H. 2005. Optically transparent bionanofiber composites with low sensitivity to refractive index of the polymer matrix. *Applied Physics Letters, 87*(24), 243110.

Oliveira, R. L., Vieira, J. G., Barud, H. S., Assunção, R., Filho, G. R., Ribeiro, S. J., & Messadeqq, Y. 2015. Synthesis and characterization of methylcellulose produced from bacterial cellulose under heterogeneous condition. *Journal of the Brazilian Chemical Society, 26*, 1861–1870.

Paâ, N., Zahan, K. A., Muhamad, I. I., & Seng, K. F. 2013. Modified fermentation for production of bacterial cellulose/polyaniline as conductive biopolymer material. *Jurnal Teknologi, 62*(2).

Pei, S., & Cheng, H. M. 2012. The reduction of graphene oxide. *Carbon, 50*(9), 3210–3228.

Poddar, M. K., & Dikshit, P. K. 2021. Recent development in bacterial cellulose production and synthesis of cellulose based conductive polymer nanocomposites. *Nano Select, 2*(9), 1605–1628.

Pourjavaher, S., Almasi, H., Meshkini, S., Pirsa, S., & Parandi, E. 2017. Development of a colorimetric pH indicator based on bacterial cellulose nanofibers and red cabbage (Brassica oleraceae) extract. *Carbohydrate Polymers, 156*, 193–201.

Qian, L., & Hinestroza, J. P. 2004. Application of nanotechnology for high performance textiles. *Journal of Textile and Apparel, Technology and Management, 4*(1), 1–7.

Ramírez, J. A. Á., Hoyos, C. G., Arroyo, S., Cerrutti, P., & Foresti, M. L. 2016. Acetylation of bacterial cellulose catalyzed by citric acid: Use of reaction conditions for tailoring the esterification extent. *Carbohydrate Polymers, 153*, 686–695.

Sahu, D., Sutar, H., Senapati, P., Murmu, R., & Roy, D. 2021. Graphene, graphene-derivatives and composites: Fundamentals, synthesis approaches to applications. *Journal of Composites Science, 5*(7), 181.

Sanyang, M. L., & Jawaid, M. (Eds.). 2019. *Bio-based polymers and nanocomposites: Preparation, processing, properties & performance*. Springer.

Saska, S., Barud, H. S., Gaspar, A. M. M., Marchetto, R., Ribeiro, S. J. L., & Messaddeq, Y. 2011. Bacterial cellulose-hydroxyapatite nanocomposites for bone regeneration. *International Journal of Biomaterials, 2011.*

Shah, J., & Malcolm Brown, R. 2005. Towards electronic paper displays made from microbial cellulose. *Applied Microbiology and Biotechnology, 66*(4), 352–355.

Shah, N., Ul-Islam, M., Khattak, W. A., & Park, J. K. 2013. Overview of bacterial cellulose composites: A multipurpose advanced material. *Carbohydrate Polymers, 98*(2), 1585–1598.

Shao, W., Wang, S., Liu, H., Wu, J., Zhang, R., Min, H., & Huang, M. 2016. Preparation of bacterial cellulose/graphene nanosheets composite films with enhanced mechanical performances. *Carbohydrate Polymers, 138*, 166–171.

Sun, B., Yuan, Z., Li, K., Tang, W., Xu, J., Zheng, H., . . . Liu, Y. 2022. Remarkable toughness reinforcement of elastomer with cellulose microgel: Employing polymer matrix and cellulose nanofiber network as the continuous phase and energy dissipation agent. *Industrial Crops and Products, 186*, 115227.

Sureshkumar, M., Siswanto, D. Y., & Lee, C. K. 2010. Magnetic antimicrobial nanocomposite based on bacterial cellulose and silver nanoparticles. *Journal of Materials Chemistry, 20*, 6948–6955.

Swingler, S., Gupta, A., Gibson, H., Kowalczuk, M., Heaselgrave, W., & Radecka, I. 2021. Recent advances and applications of bacterial cellulose in biomedicine. *Polymers, 13*(3), 412.

Thomas, B., Raj, M. C., Joy, J., Moores, A., Drisko, G. L., & Sanchez, C. 2018. Nanocellulose, a versatile green platform: From biosources to materials and their applications. *Chemical Reviews, 118*(24), 11575–11625.

Tomé, L. C., Brandao, L., Mendes, A. M., Silvestre, A. J., Neto, C. P., Gandini, A., . . . Marrucho, I. M. 2010. Preparation and characterization of bacterial cellulose membranes with tailored surface and barrier properties. *Cellulose, 17*(6), 1203–1211.

Torres, F. G., Arroyo, J. J., & Troncoso, O. P. 2019. Bacterial cellulose nanocomposites: An all-nano type of material. *Materials Science and Engineering: C, 98*, 1277–1293.

Ul-Islam, M., Khan, S., Ullah, M. W., & Park, J. K. 2015. Bacterial cellulose composites: Synthetic strategies and multiple applications in bio-medical and electro-conductive fields. *Biotechnology Journal, 10*(12), 1847–1861.

Ul-Islam, M., Khan, T., & Park, J. K. 2012. Nanoreinforced bacterial cellulose: Montmorillonite composites for biomedical applications. *Carbohydrate Polymers, 89*(4), 1189–1197.

Ul-Islam, M., Khattak, W. A., Ullah, M. W., Khan, S., & Park, J. K. 2014. Synthesis of regenerated bacterial cellulose-zinc oxide nanocomposite films for biomedical applications. *Cellulose, 21*(1), 433–447.

Ul-Islam, M., Shah, N., Ha, J. H., & Park, J. K. 2011. Effect of chitosan penetration on physico-chemical and mechanical properties of bacterial cellulose. *Korean Journal of Chemical Engineering, 28*(8), 1736–1743.

Wahid, F., Huang, L. H., Zhao, X. Q., Li, W. C., Wang, Y. Y., Jia, S. R., & Zhong, C. 2021. Bacterial cellulose and its potential for biomedical applications. *Biotechnology Advances, 53*, 107856.

Wang, J., Gao, C., Zhang, Y., & Wan, Y. 2010. Preparation and in vitro characterization of BC/PVA hydrogel composite for its potential use as artificial cornea biomaterial. *Materials Science and Engineering: C, 30*(1), 214–218.

Wang, J., Lu, X., Ng, P. F., Lee, K. I., Fei, B., Xin, J. H., & Wu, J. Y. 2015. Polyethylenimine coated bacterial cellulose nanofiber membrane and application as adsorbent and catalyst. *Journal of Colloid and Interface Science, 440*, 32–38.

Wang, X., Xie, Y., Ge, H., Chen, L., Wang, J., Zhang, S., . . . Feng, X. 2018. Physical properties and antioxidant capacity of chitosan/epigallocatechin-3-gallate films reinforced with nano-bacterial cellulose. *Carbohydrate Polymers, 179*, 207–220.

Yang, J., Sun, D., Li, J., Yang, X., Yu, J., Hao, Q., . . . Gu, J. 2009. In situ deposition of platinum nanoparticles on bacterial cellulose membranes and evaluation of PEM fuel cell performance. *Electrochimica Acta, 54*(26), 6300–6305.

Yano, H., Sugiyama, J., Nakagaito, A. N., Nogi, M., Matsuura, T., Hikita, M., & Handa, K. 2005. Optically transparent composites reinforced with networks of bacterial nanofibers. *Advanced Materials, 17*(2), 153–155.

Yano, S., Maeda, H., Nakajima, M., Hagiwara, T., & Sawaguchi, T. 2008. Preparation and mechanical properties of bacterial cellulose nanocomposites loaded with silica nanoparticles. *Cellulose, 15*(1), 111–120.

Yoon, S. H., Jin, H. J., Kook, M. C., & Pyun, Y. R. 2006. Electrically conductive bacterial cellulose by incorporation of carbon nanotubes. *Biomacromolecules, 7*(4), 1280–1284.

Zhang, F., Tang, Y., Yang, Y., Zhang, X., & Lee, C. S. 2016. In-situ assembly of three-dimensional MoS_2 nanoleaves/carbon nanofiber composites derived from bacterial cellulose as flexible and binder-free anodes for enhanced lithium-ion batteries. *Electrochimica Acta*, *211*, 404–410.

Zhang, T., Wang, W., Zhang, D., Zhang, X., Ma, Y., Zhou, Y., & Qi, L. 2010. Biotemplated synthesis of gold nanoparticle: Bacteria cellulose nanofiber nanocomposites and their application in biosensing. *Advanced Functional Materials*, *20*(7), 1152–1160.

Zhang, Z., Yang, Y., Guo, W., Chang, G., & Li, J. 2022. Synergistic capture and conversion of soluble polysulfides in Li—S batteries with composite freestanding carbonaceous interlayers. *ACS Applied Materials & Interfaces*, *14*(7), 9231–9241.

Zhang, Z., Zhang, J., Zhao, X., & Yang, F. 2015. Core-sheath structured porous carbon nanofiber composite anode material derived from bacterial cellulose/polypyrrole as an anode for sodium-ion batteries. *Carbon*, *95*, 552–559.

Zhijiang, C., & Guang, Y. 2011. Bacterial cellulose/collagen composite: Characterization and first evaluation of cytocompatibility. *Journal of Applied Polymer Science*, *120*(5), 2938–2944.

Zhou, T., Chen, D., Jiu, J., Nge, T. T., Sugahara, T., Nagao, S., . . . Xiong, D. 2013. Electrically conductive bacterial cellulose composite membranes produced by the incorporation of graphite nanoplatelets in pristine bacterial cellulose membranes. *Express Polymer Letters*, *7*(9).

Zimmermann, K. A., LeBlanc, J. M., Sheets, K. T., Fox, R. W., & Gatenholm, P. 2011. Biomimetic design of a bacterial cellulose/hydroxyapatite nanocomposite for bone healing applications. *Materials Science and Engineering: C*, *31*(1), 43–49.

10 Application of Bacterial Cellulose in the Food Industry

Yogita V. Chavan
MIT Art, Design and Technology University
Pune, Maharashtra, India

Parmjit S. Panesar
Sant Longowal Institute of Engineering and Technology
Longowal, Punjab, India

Vikas Kumar
Punjab Agricultural University
Ludhiana, Punjab, India

10.1 INTRODUCTION

Bacterial cellulose (BC) is an ecofriendly and exocellular material made from different bacterial strains, including *Komagataeibacter* (formerly *Gluconacetobacter*), *Pseudomonas, Azotobacter, Achromobacter, Aerobacter, Alcaligenes, Agrobacterium, Rhizobium, Dickeya, Sarcina, and Rhodobacter* (Ul-Islam, 2016a, 2016b; Reis, 2019; Skvortsova, 2019). In 1886, BC was reported for the first time by R. M. Brown; its structural properties and chemical composition were evaluated later in 1931 by Hibbert and Barsha. Like plant-derived cellulose, BC possesses the same properties reported by many of the previous scientists. It has the same chemical structure as plant cellulose $(C_{12}H_{20}O_{10})n$; however, BC does not contain free lignin and pectin, which is naturally present in plant-derived cellulose. Therefore, purification and clarification of BC are easier and require less energy consumption (Huang et al., 2014). In contrast, plant-derived cellulose usually requires harsh chemical treatment for purification due to the presence of lignin, hemicellulose, etc. BC possesses high water holding capacity, higher degree of polymerization, higher specific area, and longer drying time compared to plant cellulose (Sulaeva et al., 2015; Meftahi et al., 2009). It also has excellent tensile strength because of its web-like and ultrafine network structure, high purity, high degree of porosity, relatively high permeability to liquid and gases, which make it suitable for various applications in different fields. BC has been exploited in many industries, among them medical engineering, electronics, biosensors, paper making, cosmetics, and food and food packaging (Ullah et al., 2016). Due to its excellent and multifarious properties, BC has ample scope for future research.

10.2 PROPERTIES OF BACTERIAL CELLULOSE

Cellulose is a plant-derived polymer, the most abundant and viable, sustainable biopolymer available and predominantly utilized on the earth. To find alternative sources for cellulose, efforts has been made to develop fermentation techniques for the production of BC using different bacterial genera. *Glucanobacter* is most likely to be used for this purpose. The chemical formula of BC is comparable to that of plant-based cellulose, but it has different physical and chemical properties (Chau et al., 2008). The degree of polymerization varies according to the material, from 2000–6000 in BC to 13,000–14,000 in plant cellulose. BC's three-dimensional structure guards against

DOI: 10.1201/9781003355434-10

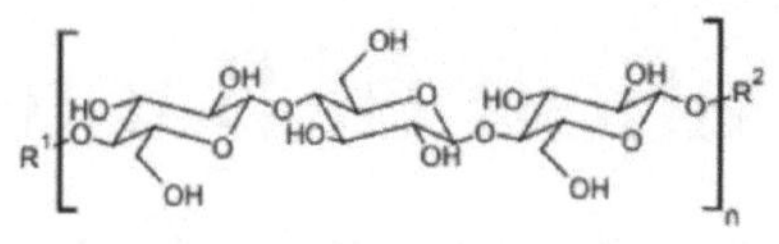

FIGURE 10.1 Chemical structure of cellulose.

microbial invasion. The free surface hydroxyl group on the cellulose is a result of the glycosidic bond (β-1–4). BC measures 10–50 nm in diameter and 100–1000 nm in length. (Czaja et al., 2007; Petersen, 2011). As a result of the abundance of oxygen atoms and hydroxyl groups, it has a lot of hydrogen bonds, shown in Figure 10.1 (Pogorelova et al., 2020). Van der Waals forces, which aid in the formation of crystalline nanofibers and then the progress of microfibrillar structure, cause parallel stacking to be seen in cellulose (Chen et al., 2018). The presence of a supercoil structure in BC aids in fostering the hierarchical orders that give cellulose a very high level of mechanical strength (Klemm et al., 2018).

10.3 MICROBIAL SOURCES OF BACTERIAL CELLULOSE

BC can be synthesized by microbial fermentation using various species of bacteria. Among bacterial species, Gram-negative and rod-shaped aerobic bacteria, for example, *Gluconacetobacter xylinus* (*Acetobacter xylinum*) are widely preferred for production (Zhong, 2020), because of their capability to produce higher cellulose yield. It is considered to be the index organism for the production of BC at the commercial level (Keshk, 2014). This bacterial group is known to produce microfibrilar cellulose on a large scale (Ross et al., 1991; Tanaka et al., 2000). Apart from this group, several other bacterial species also have been utilized for the production of BC, mainly *Azotobacter, Rhizobium* spp., *Acetobacter* spp., *Sarcina*, *Agrobacterium* spp., *Alcaligenes*, and *Pseudomonas* (Jonas and Farah, 1998; Panesar et al., 2009; Panesar et al., 2012). *Acetobacter,* common vinegar-producing bacteria, are purple in color and nonphotosynthetic, and transform glucose, glycerol, other sugars, and any other organic materials into pure cellulose (Brown and Montezinos, 1976; Son et al., 2001). BC microfibrils merge in the growth medium to create large-ribboned cellulose, which is immobile because it forms floating pellicles with the entangled associated cells. *Agrobacterium tumefaciens,* tumor-forming bacteria, also secrete cellulose fibrils around the cell to encourage cell attachment and virulence (Matthysse et al., 1995). In a suitable medium, both *Agrobacterium tumefaciens* and *Gluconacetobacter xylinus* can grow widely, but mutants are less capable of cellulose synthesis. Additionally, it has been discovered that the genus *Komagataeibacter* is remarkable for its capacity to produce cellulose (Wang et al., 2018).

BC has been examined for its possible application in many areas, including food and packaging industries, in which it can be used as a rheology modifier (thickening, gelling, and stabilizing agent), low-calorie dessert or other low-calorie product preparation, fat replacer, and as an active ingredient or delivery medium for the release of active compounds in food packaging (Akoglu et al., 2018; Bandyopadhyay et al., 2020; Jayani et al., 2020; Razavi et al., 2020). It can be used as a flexible matrix for developing biomaterials in the electronics, electrochemical, and magnetic fields (Badshah et al., 2018; Beekmann et al., 2020). Due to its liquid holding capacity and control release ability, it has been used in drug delivery systems (Inoue et al., 2020; Luo et al., 2017; Weyell et al., 2019). In addition to its pain-reducing ability, by maintaining moisture content it can act as a bacteriological barrier in dressings for the treatment of severe burns and chronic wounds. Halib et al. (2019), Frone et al. (2020), and Zhang et al. (2020a) have also reported its use in tissue engineering. The morphological structure of BC lends itself to superior biocompatibility and high tensile strength, allowing it to be used to assist newly formed tissues, interact with cells, and release compounds essential for cell growth. In a number of applications, BC can be utilized in biosensors to identify and measure specific compounds in the product. Cell and enzyme immobilization with BC has shown promising results in the biomedical field, food manufacturing processes, and effluent treatment (Hu et al., 2011; Moradi et al., 2019; Cai et al., 2018). The use of BC in combination with specific materials for water filtration to eliminate contaminants, heavy metal components, and water impurities has been reported by Urbina et al. (2018), Zhuang and Wang (2019), Nunez et al. (2020), Shoukat et al. (2019), and Luo et al. (2020). As discussed earlier, due to its ability to control the release of substances, it can stabilize emulsions and has been used to produce cosmetic products (Wang et al., 2020; Amnuaikit et al., 2011; Aramwit and Bang, 2014).

10.4 APPLICATIONS OF BACTERIAL CELLULOSE IN FOOD PROCESSING INDUSTRY

The US Food and Drug Administration (FDA) has approved BC (a dietary fiber) as a "generally recognized as safe" (GRAS) food (Shi et al., 2014). Due to its indigestibility by humans, BC has been used for dietetic foods (Mohite and Patil, 2014). Like other dietary fibers, it facilitates intestinal transport and also improves the mouthfeel (Fontana et al., 2017). Due to its multifunctional characteristics, BC has gained importance in various areas and has huge scope for commercialization (Table 10.1). This chapter focuses mainly on its applications in the food and food packaging industry. Its key applications in the food industry are described in Table 10.1 and represented graphically in Figure 10.2.

TABLE 10.1
Potential Applications of Bacterial Cellulose (BC) in the Food Industry

BC in Food and Related Products	Type/Form of BC	Significance of BC Addition	References
Dessert (nata de coco)	Slices/cubes	Main structural component	Iguchi et al. (2000)
Beverages			
Chocolate drink	Liquid paste	Stability against heat	Okiyama et al. (1993)
Wine (beverage)	Pieces	Solid support to increase activity of yeast	Montealegre et al. (2012), Nguyen et al. (2009), Ton and Le (2011), Trevino-Garza et al. (2020)
Food beverages (yoghurt, milk, fruit drinks)	Powder	Thickener and stabilizer	Dourado et al. (2016)
Semiliquid and Solid Food Products			
Vegetarian meat/artificial	Formulated sheets	Structural component and fat adsorbent	Juzlova et al. (1996), Purwadaria et al. (2010), Wonganu and Kongruang (2010)
Low-cholesterol diet food	Powder	Fat adsorbent	Chau et al. (2008), Lin and Lin (2004), Stephens et al. (1990)
Pasty food and jams	Aqueous paste	Heat-stable suspending and bulk forming agent	Okiyama et al. (1992)
Tofu and kamboko	Aqueous paste	Gelling and hardening agent, texture modifier	Okiyama et al. (1993)
Ice cream mix	Aqueous paste	Improved stability against freeze-thaw process, having good textural properties	Okiyama et al. (1993) Guo et al. (2018)

(Continued)

TABLE 10.1 (*Continued*)
Potential Applications of Bacterial Cellulose (BC) in the Food Industry

BC in Food and Related Products	Type/Form of BC	Significance of BC Addition	References
Enzyme Immobilization			
Enzyme glucoamylase	Beads	Support to increase enzymatic activity	Wu and Lia (2008), Wu et al. (2013)
Fungal laccase	Sponge	Support to increase activity of laccase enzyme	Chen et al. (2015)
L-lysine	Cubes	Solid support to increase activity and cell viability	
Food Packaging Development			
Biodegradable and bioactive film	BC-active component	To improve the shelf life and maintain the quality	Gedarawatte (2020), Ma et al. (2020), Shafipour et al. (2020)
Food packaging	Sheets, film, and powder	Hydrophobic and antimicrobial packaging	Dobre et al. (2012), Jipa et al. (2012), Tomé et al. (2010)
Active packaging film	Powder	Improve tensile strength	Xie et al. (2020)
BC-nanoparticle composites	Powder, sheet	Antimicrobial agent to improve product quality	Jebel and Almasi (2016), Dobre and Stoica-Guzun (2013), Khan et al. (2015)

Source: Adapted from Ullah et al., 2016.

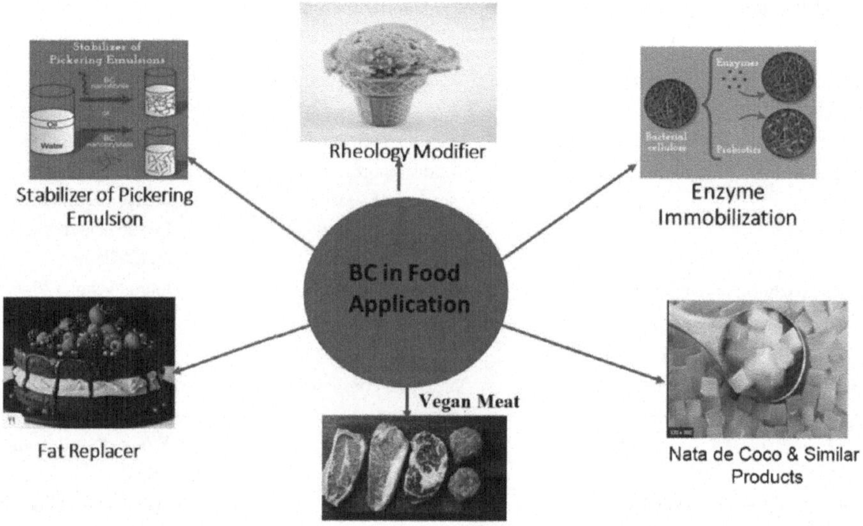

FIGURE 10.2 Main applications of bacterial cellulose (BC) in the food industry.

10.4.1 Dessert Preparation

BC is typically used as a base for chewy, juicier nata de coco dessert. The product has a gel-like consistency, contains no cholesterol, and has low fat and calorie content. Nata de coco originated in the Philippines and is a traditional dessert in southeast Asian countries (Iguchi et al., 2000). It is made by cultivating BC in coconut water with the addition of carbohydrates and amino acids to improve the production environment. The thick sheet of cellulose produced is then further processed by cutting, washing, and cooking in sugar syrup for various food product applications, such as jellies, fruit cocktails, and desserts (Figure 10.2). Supplementation of nata de coco with cereals was proved to be a high dietary food, which reduces consumers fat level after consumption (Mesomya et al., 2002). Its smooth texture, tempting mouthfeel, and easy manufacturing process makes it popular in Asia and it has spread rapidly throughout the world.

BC can be produced by altering fermentation parameters and varying the formulation of the culture medium to make it available in different shapes, sizes, and textures. Also, the culture medium can be enriched by incorporating different flavors to fulfill consumer requirements and demand. Therefore, efforts have been made to produce similar kinds of product by adopting new processes and formulation. According to Iguchi et al. (2000) and Jozala et al. (2016), culture medium resources such as nata-de-pina, which is made from pineapple flavor, can be used to control the product's flavor. As a result, BC has untapped potential as a raw material for a variety of novel products, such as low-cholesterol products, snacks, and low-calorie sweeteners, currently in high demand.

10.4.2 Rheology Modifier

According to the scientific literature, BC's structure and properties make it appropriate for an extensive range of uses in the food processing sector. Okiyama et al. (1992, 1993) investigated the functional properties of BC with respect to rheology modification and the study demonstrated its potential benefits as a gelling, stabilizing, and thickening agent and as a water-binding agent in several food products. As mentioned, BC can be added at lower concentrations as a raw ingredient to a variety of food formulations to prevent flavor interaction during processing and to improve product stability under various storage conditions. The uses of BC as rheology modifier are categorized and discussed in detail in the following sections.

10.4.2.1 Thickening/Surfactant Agent

BC has been used as a functional food additive in the food processing industry. The major category of application is in liquid foods and beverages such as milk beverages. For example, soybean milk, coffee, and porridge require suspension during storage. Xanthan gum, pectin, carboxymethylcellulose (CMC), and soybean polysaccharide are known for their use as thickeners and surfactants in liquid food products to achieve uniform particle suspension (Dourado et al., 2016). The addition of these thickeners, however, typically results in poor suspension stability, transparency interferences, and phase separation (Swazey and Madison, 2010). Moreover, viscous food products give an unpleasant taste to consumers, and therefore a novel suspension agent is needed to provide the required dispersion stability and viscosity. With this perspective in mind, BC pellets were made through an agitated fermentation process and were discovered to have excellent suspension properties (San-Ei Gen, 2020). The addition of 0.2% to 0.3% of BC improved the gel strength and textural properties of tofu (a product prepared from coagulating and pressing soy milk), as reported by Okiyama et al. (1993). The addition of BC into pasty condiments was shown to improve the gel strength by reducing the stickiness of the product and making it easy to serve using a spoon (Shi et al., 2014). The addition of BC to kamaboco, a processed Japanese seafood, eliminates the springiness and improves the product's stiffness and brittleness, which satisfies consumer demand (Okiyama et al., 1993). Also, the addition of BC to chocolate drink protected it from cocoa precipitation due to the retention properties of BC. It has better stability during the heat sterilization process and its viscosity does not change (Okiyama et al., 1993). BC was found to be a cheaper alternative source of thickening agent compared with traditional material (xanthan gum and locust bean gum). The unique nanofiber-weaved, three-dimensional reticulated network of BC confers its excellent suspension capability of insoluble particulates with low viscosity. BC performs its functional properties after addition of it in any solution, it suspends well and forms a suspension even at lower concentrations. Thus, BC has proven its well suspension ability rather than solubilization in solution without altering its charge.

10.4.2.2 Stabilizing/Emulsifying Agent

The primary purpose of a stabilizer is to increase the stability of solutions by adhering to the immiscible emulsion interfaces and stabilizing droplets by forming a robust monolayer at the interface to prevent amalgamation. This type of emulsion is generally known as a pickering emulsion (Figure 10.2; Hu et al., 2015). When compared to commercial cellulose derivatives, the addition of BC demonstrated improved stability in an oil-water emulsion containing 10% olive oil (hydroxypropyl methylcellulose (HPMC) and CMC). These results indicated that BC formed a robust fibril network that readily adsorbs to the oil phase/interphase (Paximada et al., 2016b). Even though the BC-added emulsion had larger droplet size (26 m), it was the most stable to coalescence and had a 3% serum index, which was the lowest of the substances tested. In a different study, the same team used olive oil at 5% by weight, BC at 0 to 1% by weight as a stabilizer, and whey protein isolate at 2% to 5% by weight as a surfactant. The result suggests that the emulsion was unstable at 0.5% to 0.7 wt%; beyond this concentration BC promotes stable emulsions at 1 wt%, which avoids amalgamation (Paximada et al., 2016a). Likewise, ice cream is globally appreciated and consumed as a frozen dessert, but upon temperature fluctuation the partially united fat globules collapse due to the melting of ice crystals (Petrut et al., 2016). To avoid this loss, different plant-based hydrocolloids have been used to increase consumer acceptability by retaining the overall properties of the product. The addition of CMC has long been explored as a stabilizer in ice cream mix to hold the water. Increased viscosity prevents melting while reducing the formation of ice and lactose crystals in ice cream mixtures during storage has been reported by Granger et al. (2005) and Guo et al. (2018). Despite its superior stabilizing ability compared with plant cellulose, its in ice cream remains to be explored (Paximada et al., 2016a). The addition of BC to the ice cream mix allowed it to maintain its texture for at least 60 minutes at room conditions, whereas ice cream without BC melted completely over the same amount of time. Another combination is BC/soy protein isolate composites, which were reported to improve the textural, rheological, emulsifying properties and thermal stability of ice cream (Guo et al., 2018). In addition, BC has also been proven to have the potential for application in a number of products as a suspended and emulsion stabilizer for olive oil (Yan et al., 2017), edible foam (Zhang et al., 2020b), cheese, mayonnaise (60), and cakes (Karahan, 2011; Rios et al., 2018).

10.4.2.3 Fat Replacer/Low-Cholesterol Product

Fat is an ingredient that plays a significant role in maintaining the texture and physical properties of a variety of food products (Table 10.2). The primary purpose of substituting BC for fat to imitate its functionality in a variety of products is to lower the calorific value of the product to prevent associated health problems. Fat can be reduced by preparing a double emulsion, a multilayer system that includes different phases. According to Tekin et al. (2017), emulsions such as oil-in-water and water-in-oil are single-phase emulsions dispersed into a third phase, whereas water-in-oil-in-water and oil-in-water-in-oil are double emulsions. Several reports are available on the application of BC as a fat replacer in many products. Compared to a single-phase emulsion, designing low-fat products with double emulsion provides improved mouthfeel that replicates the satisfying texture of nonfat or low-fat incorporated products. Studies have also been done on the use of plant-based cellulose for different applications as a bulking agent, texturizer, and a natural low-fat or fat replacer in a variety of products, such as emulsified meat products (Kumar et al., 2016), cheese (Aydinol and Ozcan, 2018), and cakes (Rios et al., 2018).

Interest in exopolysaccharides has increased because of their better purity and versatile properties, making them a suitable ingredient for many food products. Exopolysaccharides have been utilized as a low-calorie food ingredient and fat replacer in many animal-originating products, such

TABLE 10.2
Potential Applications of Bacterial Cellulose (BC) as Fat Replacer in Animal Food Products

Sr. No.	Meat/Seafood Products	Technological Aspects and Quality Attributes	References
1	Chinese meatballs	After addition of different concentrations (10%, 20%, and 30%), humidity, cooking loss, water activity, juiciness and water holding capacity of product increases whereas textural properties, TBARs, springiness, firmness, shear stress of meat ball decreases. Other parameters such as chewiness, emulsion stability, and overall acceptability is like the control.	Lin and Lin (2004)
2	Dolphin-fish surimi	Increased shear stress and water holding capacity of product were increased when added to 5%, 10%, 15%, and 20% BC, whereas decrease in hardness was observed. All other properties are similar to the control.	Lin et al. (2011)
3	Frankfurter sausage	Sausages were developed using 10%, 20%, and 30% BC, with increases in product firmness, maintaining fat content and color value (L*) of product. Overall acceptability, juiciness, cooking yields, color values (b*) similar to the control.	Yu and Lin (2014)
4	Dry, fermented, and spicy sausages	Increases humidity, overall acceptability, and color attributes (a*) after addition of 5%, 10%, 15%, and 20% BC. Decreases fat, textural properties, and color value (L*). Color value (b*) is similar to the control sample.	Akoglu et al. (2018)
5	Sausages	Increases water content, textural property (hardness), water holding capacity, and cohesiveness (0.13%, 0.2%, 0.27%, 0.40%, and 0.53% BC). Decreases water activity, springiness, color values (L*, a*), resilience, adhesiveness. Other properties are similar to the control: b*, process yield.	Marchetti et al. (2017)
6	Emulsified meat products	BC maintained the quality attributes of the emulsified meat compared to the control; and proved to be a potential fat replacer.	Oliveira et al. (2021)
7	Meat emulsion	BC is used as a fat substitute, interacting with non-meat ingredients, affecting rheological parameters, emulsion stability, and water and fat holding capacities.	Kumar (2021); Marchetti and Andrés (2021)

TBARs: thiobarbituric acid reactive substances; ES: emulsion stability.

Source: Adapted from Oliveira et al. (2021).

as meat products (Marchetti, 2017), in meatballs (Lin and Lin, 2004), surimi products (Lin et al., 2011), cheese (Karahan et al., 2011), ice cream (Guo et al., 2018), whey protein isolate, and mayonnaise (Akoglu et al., 2018). Lin and Lin (2004) studied the effect of the addition of 10% BC to meatball formulation and observed similar properties to the control sample. The BC-fortified meatballs were sensorially acceptable, shelf stable, juicy and chewy, therefore BC can be utilized to replace fat in different emulsified meat products. Okiyama (1993) suggested its application in different processed products such as surimi, sausages, and hamburger patties. It was observed that hardness was increased with a reduction in springiness and gel strength. Frankfurter-type sausages developed using 10, 20, and 30% concentration of BC showed reduced losses during cooking compared to a control formulation (Yu and Lin, 2014). BC added to low sodium and low fat content sausages showed good cooking performance and therefore it is acceptable as a fat replacer in different types of food products (Marchetti et al., 2017; Lin et al., 2011). The volume, softness, and flavor of the product, especially cakes, are further enhanced by the presence of fats which maintain air cells in the emulsion phase (Rios et al., 2018). Guo et al. (2018) used various BC/SPI (soy protein isolate) ratios from 1:20 to 1:5 as a partial or full fat replacer to investigate the effect of BC/SPI composites on an ice cream model. They reported that, by adding 20% BC/SPI (1:20) to the formulation, which results in a texture that is similar to the usual ice cream mix with 30% cream addition, product-specific properties can be improved.

Purified BC has a cholesterol-lowering effect when added to food products (Stephens et al., 1990). According to an in vivo study, adding BC to hamster diets lowers liver cholesterol, serum triglycerides, and total cholesterol level (Chau et al., 2008). The comparative studies were carried out by feeding plant cellulose and BC to rat, which enhanced the defecation of total lipids (144% to 182%), cholesterol (136% to 203%), and bile acids (259% to 479%) (Chau et al., 2008).

According to reports, BC is a low-calorie bulking ingredient that has been used to produce a variety of novel functional foods (rich in fiber) in a variety of forms or formulae. BC can be used to stop obesity in mice that has been brought on by high-fat diet consumption. The addition of BC protects against liver damage, inflammation, and insulin resistance, while also improving antioxidant defence and controlling the release of adipocytokines and proteins linked to adipogenesis (Zhai et al., 2018a). Furthermore, BC protects colonic muscle cells against apoptosis because supplementation with BC effectively increases the length and thickness of colonic mucosal muscle cell. From the literature it is revealed that BC has a great potential as a dietary fiber if used in food products (Zhai et al., 2018b, 2018c). An excellent in vitro model was developed by Mikkelsen et al. (2009), who incorporated soluble plant cell wall polysaccharide with BC to study the effect of the composite on plant cell wall dietary fiber.

10.4.3 Vegan Meat/Vegetarian Meat/Artificial Meat

The literature suggests the use of BC with *Monascus* extract (a natural pigmented mold) is applicable for the production of vegetarian meat analogs (Sheu et al., 2000; Purwadaria et al., 2010). *Monascus purpureus*, a naturally pigmented mold, has anti-hypercholesterolemic and cholesterol-lowering effects due to the presence of monacolin and mevinolin (Kim and Ku, 2018). During fermentation, this mold produces red, yellow and orange pigmented BC products. The production of colored pigments depends on the selection of carbon and nitrogen sources in the fermentation medium (Sheu et al., 2000). The developed color composites are resistant to washing and other processing techniques (Sheu et al., 2000). This composite can be utilized as a starting point for the creation of novel functional foods, most notably artificial meat or vegan meat, as shown in Figure 10.2 (Ochaikul et al., 2006; Purwadaria et al., 2010). The morphology and color of the *Monascus*-BC complex are consistent with what has been said in the discussion so far (Purwadaria et al., 2010; Wonganu and Kongruang, 2010). The complex's flavor is perfect for substituting vegetarian meat for meat and fish. Due to its versatile properties, the *Monascus*-BC complex has market appeal and is in high demand. It will be the non-animal-origin product of the future for the vegetarian consumer and even nonvegetarian consumers with dietary restrictions. Also its cholesterol-lowering effect make it suitable for substitution (Ng et al., 2004; Ng and Shyu, 2004; Ullah et al., 2016).

10.4.4 Enzyme and Cell Immobilization

By altering operating parameters, the technology of enzyme immobilization supports food production in the food and fermentation industries (Fernandes, 2010. In recent times cellulosic biomaterials have taken on more significance in enzyme immobilization technology (Koutinas et al., 2012). This is due to the unique material qualities of BC; its pure form is currently receiving more attention as a new useful biomaterial than plant-based cellulose (Petersen and Gatenholm, 2011). By immobilizing the glucoamylase enzyme on BC beads, it can be used in the food business (Wu and Lia, 2008). According to Wu, Lia, and Ho, the stability of BC-immobilized enzymes can be increased at lower pH and temperature values with the lowest bead size (0.5 to 1.5 mm) (2013). Another attempt was made to physically adsorb the laccase enzyme (fungus) on the BC and cross-link it with glutaraldehyde. When compared to a free and adsorbed laccase enzyme, it exhibits good catalytic activity, provides superior stability to immobilized laccase enzyme, and maintains 69% of its initial activity after numerous cycles (Chen et al., 2015). Many additional enzymes, such as laccase for bioanalysis and horse radish peroxidase for biosensors, have been reported to be immobilized by BC (Lv et al., 2016; Wang et al., 2010; Zhang et al., 2010).

Additionally, the immobilization procedure, which uses yeast that has been immobilized on BC for batch fermentation, has been employed in the wine industry. This method is inexpensive and lowers the costs associated with the manufacturing of inoculum for the wine producer. Moreover, at the conclusion of the fermentation process, the utilized yeast can be recovered via separation, as reported by Nguyen et al. (2009). The experiments were performed to check the effect of calcium alginate (CA) alone or a BC/CA composite on ethanol yield. The results showed that yeast immobilization on BC/CA composites can increase ethanol yield compared with yeast immobilized on CA alone (Kirdponpattara and Phisalaphong, 2013). But in another study no significant difference in ethanol yield was observed between BC-immobilized yeast and CA-immobilized culture (Ton and Le, 2011). The effective ethanol production as a result of the superior characteristics of BC/CA composites. It offers great thermal, mechanical, porosity, and hydrophilicity stability (Montealegre et al., 2012).

Likewise, L-lysine can be synthesized by immobilizing *Corynebacterium glutamicum* on BC (Kirdponpattara and Phisalaphong, 2013). Lysine yield can reach up to 95% using immobilized cells during fermentation. The cell has very good stability, cell viability in sterile water (pH 7) and can be stored at 40°C for 30 days (Tam and Huong, 2014). The study revealed that BC has good potential for immobilizing enzymes and cell systems to improve the stability and activity of enzymes and cells, resulting in an improvement in food industry yield, quality, and stability of products. The potential of immobilized probiotics and enzyme probiotics to balance gut microbiota, which has positive effects on health, has raised demand for these products. Probiotics typically have a limited chance of surviving while being stored and moving through the gastrointestinal tract (Chavarri et al., 2012). *Lactobacillus* spp., a probiotic strain that has been immobilized by integrating it into either BC membranes or the culture medium used to produce BC. In terms of defending the efficiency of probiotic from the effects of gastric juices and bile salts, the second technique was the most successful. However, this method is inconsistent to purify BC membranes after fermentation, which entails soaking at temperatures often greater than 100°C in an alkaline solution. As a result, BC serves as a helpful support for immobilizing *Lactobacillus* cells and shielding them from stomach fluids.

Unfortunately, investigations on BC utilized probiotic-containing materials typically only consider how technological factors affect bacterial survival rather than how well BC can release probiotic bacteria in the colon to fulfil their function. The effectiveness of any product is reliant on the bacteria being released at their invasion sites, so future studies are needed to address such difficulties. In addition to that, lysozyme, laccase, and lipase have all been effectively immobilized for sustained release from BC (Wu et al., 2017; Chen et al., 2015; Bayazidi et al., 2018).

10.4.5 Food Packaging

The most thoroughly researched field of BC in the literature relates to food packaging for the creation of bioactive films and biodegradable packaging. To lengthen the shelf life of food goods, BC has been added to a wide range of substances (Gedarawatte et al., 2020), for example, cotton fibers (Ma et al., 2020), postbiotics of lactic acid bacterium (Shafipour et al., 2020), and potato peel are a few of the materials to which BC has been added (Xie et al., 2020). Edible coatings or biodegradable films are membranes formed from renewable materials by incorporating bioactive ingredients for food application. To form cohesive networks, at least two components must be employed, including a macromolecule and a plasticizer able to reduce brittleness innate in most materials. Other components may also be fused to improve mechanical and barrier properties. A desirable material feature can occasionally be produced by combining various matrices.

As a food-approved material, BC matrix can be utilized to make edible packaging films and coatings for different purposes, including food packaging. Thin packaging films and coatings are usually applied on the surface of packaged products, which creates a modified atmosphere for the product, thus extending its shelf life. A variety of edible materials containing flavor components can be used for nontraditional sandwiches (Otoni et al., 2017) or for different snack products. The main purpose to develop innovative packaging material is to make films and coatings more effective at protecting them from vapor and gas than traditional packaging materials. In active packaging, different agents or scavengers, such as antimicrobial agents, oxygen, ethylene, and moisture absorbers are used to extend the storage period and enhance the safety of the product by removing undesirable material from the packaging.

The packaging material plays a vital role to be strong and retain good barrier properties. A variety of polymers have been combined with BC to achieve these properties, bringing benefits related to their specific properties. Chitosan films were created by infusing BC into a chitosan solution, taking advantage of chitosan's antimicrobial properties (Lin et al., 2013b). A composite film produced with this method was effective against all type of bacteria, and it also had a significantly higher modulus of elasticity than a neat BC film. Esterification of BC membrane was reported by Tome et al. (2010), who observed improved barrier properties compared to the original BC membrane.

The surface of vacuum-packaged frankfurters was treated with BC containing nisin to control *Listeria monocytogenes* and aerobic bacteria. It is therefore suggested that the nisin-incorporated BC active film is resistant or has a bactericidal effect on *L. monocytogenes* (Nguyen et al., 2008; Jipa et al., 2012). Polylactic acid and bacterial cellulose were combined to create a composite that has better mechanical properties for food packaging while retaining its transparency and biocompatibility. In another attempt to create active sausage casings, BC tubes were coated with the potent antibacterial material ε-polylysine (ε-PL), which resulted in lower oxygen permeability and tensile strength than the usual packaging material (Zhu et al., 2010). A BC-SA (sorbic acid) composite was developed with the ability to release the antimicrobial agent in a controlled way to improve the product shelf life by reducing microbial growth (Jipa et al., 2012). *Listeria monocytogenes* growth was shown to be more effectively controlled by bacteriocin-incorporated BC membranes than by a packaging layer with free bacteriocins, suggesting that BC served as a protective layer for the peptides (Malheiros et al., 2018). Multilayer films were used to control the antimicrobial agent's release rate, with an antimicrobial layer sandwiched between two outer layers (Jebel and Almasi, 2016).

Nanomaterial, such as silver, copper, gold, and ZnO, are known for their effective antimicrobial activity. Silver-based nanoparticles have drawn the most attention among these, due to their antimicrobial activity against a variety of bacteria, fungi, and viruses and their relative lack of toxicity to people. Several researchers invented the nanoparticle composite known as BC-silver and reported its antimicrobial activity against model microbes such as *E. coli, Staphylococcus aureus,* and *Bacillus subtilis.* The composite showed high antimicrobial activity against these organisms (Maneerung et al., 2008). An attempt was made to utilize ZnO nanoparticles in BC based packaging material. The active layer of ZnO nanoparticles with wet BC membranes were coated (on both sides) and allowed to dry. The existence of ZnO nanoparticles in packaging material provides antibacterial activity, decreases the water vapor permeability, and enhances the tensile strength of the packaging films. Other researchers have investigated the effect of TiO_2 nanoparticles as an antimicrobial agent in the development of a BC-base delivery system. Ag and ZnO nanoparticles were also explored as antimicrobial agents in BC-based active film (Dobre and Stoica-Guzun, 2013; Khan et al., 2015). To deliver the bioactive agents, BC-nisin film was designed to control *Listeria monocytogenes*. Because of its high specific surface area and gelling behavior, BC has great potential for bioactive agent delivery (Ullah et al., 2016). The result of adding L-ascorbic acid to BC aerogels confirms their suitability as a matrix for sustained release. This is because of the high rehydration property of the BC matrix, which may be associated with high pore volume (Ullah et al., 2016). When combined with serum albumin, BC has demonstrated its effectiveness as a means of delivering proteins. Studies were carried out using freeze-dried BC (FDBC) and they revealed that with FDBC, less protein is required compared to the BC produced by conventional dry procedures (Muller et al., 2013).

10.4.6 Miscellaneous Applications

Another application for BC involves replacing the fat in surimi with BC. The surimi-BC complex has a greater capacity to hold water. As a result, the surimi's network structure was enhanced by BC and maintained its original structure. The BC provided good mechanical properties. When more hydrocolloid was added to microfibrillated cellulose, cooking losses in hamburgers were decreased (Ström et al., 2013). The fat, protein, sugar, and water play important roles to protect the structural and maintain the textural properties of ice cream. By creating a three-dimensional network that surrounds the air cells that form during the whipping process and contribute to the ice cream's rich flavor, the fat enhances the product's structure and textural qualities (Varela et al., 2014). Another structural component that serves a functional role in ice cream is protein, which also helps to whip up air bubbles, emulsify fat globules during homogenization, and bind water to increase viscosity, extend melting time, and lessen iciness (Varela et al., 2014). Sugar is another important ingredient, which reduces the mixture's hardness by lowering the freezing point during processing. The addition of stabilizers and emulsifiers, which control the formation of ice crystals, results in a homogeneous product texture (Tekin et al., 2017).

10.5 CONCLUSIONS

BC has been proven as a potential alternative to plant-derived cellulose in several fields of research. Due to its unusual properties, BC has great potential applications in the food industry as a food raw material or food packaging material. Furthermore, its biodegradability, nontoxic nature, and availability from microbial sources and renewable resources (e.g., agro-industrial wastes; Singh et al., 2017) are important for environmental protection and food safety. According to the current analysis, the global microbial and bacterial cellulose production has been valued at US $506.8 million in 2020 and this is projected to reach $980.9 million by 2028, at a compound annual growth rate of 8.1% (www.reportsanddata.com/report-detail/microbial-and-bacterial-cellulose-production market). With the advent of novel technologies for BC production, such as 3D printing, the future of BC seems promising (Shaffner et al., 2017). Surface modifications in BC are possible due to the chemically reactive sites present within the BC structure (by using plasma, for instance). Surface modification in BC to improve its porosity and provide tailor-made (can be controlled by the choice of the plasma gas and other operational parameters) surface functionalities by keeping its core properties (Pertile et al., 2010). A variety of food applications have been found, such as improved fresh produce film gas permeability or higher active ingredient impregnation capability. Though validation is still necessary for the majority of food applications, commercial BC manufacture and use for low-cost niche markets remain challenging. There are reasons to think that BC has a bright future as a distinctive component for various creative food items with gastronomic and gastronomic attractions in edible films and coatings.

10.6 REFERENCES

Akoglu, A., Cakir, I., Karahan, A.G., and Cakmakci, M.L. 2018. Effects of bacterial cellulose as a fat replacer on some properties of fat-reduced mayonnaise. *Romanian Biotechnological Letters* 23:13674–13680.

Amnuaikit, T., Chusuit, T., Raknam, P., and Boonme, P. 2011. Effects of a cellulose mask synthesized by a bacterium on facial skin characteristics and user satisfaction. *Medical Devices: Evidence and Research* 4:77–81.

Aramwit, P., and Bang, N. 2014. The characteristics of bacterial nanocellulose gel releasing silk sericin for facial treatment. *BMC Biotechnology* 14(9)(December):104. http://dx.doi.org/10.1186/s12896-014-0104-x. PMid:25487808.

Aydinol, P., and Ozcan, T. 2018. Production of reduced fat Labneh cheese with inulin and β-glucan fibre based fat replacer. *International Journal of Dairy Technology* 71:362–371. https://doi.org/10.1111/1471-0307.12456.

Badshah, M., Ullah, H., Khan, A.R., Khan, S., Park, J.K., and Khan, T. 2018. Surface modification and evaluation of bacterial cellulose for drug delivery. *International Journal of Biological Macromolecules* 113:526–533. http://dx.doi.org/10.1016/j.ijbiomac.2018.02.135. PMid:2947541.

Bandyopadhyay, S., Saha, N., and Saha, P. 2020. Comparative analysis of bacterial cellulose based polymeric films for food packaging. *AIP Conference Proceedings* 2205:020069. http://dx.doi.org/10.1063/1.5142984.

Bayazidi, P., Almasi, H., and Asl, A.K. 2018. Immobilization of lysozyme on bacterial cellulose nanofibers: Characteristics, antimicrobial activity and morphological properties. *International Journal of Biologycal Macromolecule* 107:2544–2551. doi: 10.1016/j.ijbiomac.2017.10.137.

Beekmann, U., Schmölz, L., Lorkowski, S., Werz, O., Thamm, J., Fischer, D., and Kralisch, D. 2020. Process control and scale-up of modified bacterial cellulose production for tailor-made antiinflammatory drug delivery systems. *Carbohydrate Polymers* 236:116062.

Brown, R.M.J., and Montezinos, D. 1976. Cellulose microfibrils: Visualization of biosynthetic and orienting complexes in association with the plasma membrane. *Proceedings of the National Academy of Sciences of the United States of America* 73:143–147.

Cai, Q., Hu, C., Yang, N., Wang, Q., Wang, J., Pan, H., Hu, Y., and Ruan, C. 2018. Enhanced activity and stability of industrial lipases immobilized onto spherelike bacterial cellulose. *International Journal of Biological Macromolecules* 109:1174–1181.

Chau, C.F., Yang, P., Yu, C.M., and Yen, G. 2008. Investigation on the lipid-and cholesterol lowering abilities of biocellulose. *Journal of Agricultural and Food Chemistry* 56:2291–2295.

Chavarri, M., Marañon, I., and Villarán, M.C. 2012. Encapsulation technology to protect probiotic bacteria. *Probiotics*, ed. E.C. Rigobelo, 501–540. London: InTech.

Chen, G., Wu, G., Alriksson, B., Chen, L., Wang, W., Jönsson, L.J., and Hong, F.F. 2018. Scale-up of production of bacterial nanocellulose using submerged cultivation. *Journal of Chemical Technology and Biotechnology* 93:3418–3427.

Chen, L., Zou, M., and Hong, F.F. 2015. Evaluation of fungal laccase immobilized on natural nanostructured bacterial cellulose. *Frontier in Microbiology* 6:1245.

Czaja, W.K., Young, D.J., Kawecki, M., and Brown, R.M. 2007. The future prospects of microbial ccellulose in biomedical applications. *Biomacromolecules* 8:1–12.

Dobre, L.M., Stoica-Guzun, A., Stroescu, M., Jipa, I., Dobre, T., Ferdes, M., and Ciumpiliac, S. 2012. Modelling of sorbic acid diffusion through bacterial cellulose-based antimicrobial films. *Chemical Papers* 66:144–151.

Dobre, M.L., and Stoica-Guzun, A. 2013. Antimicrobial Ag-polyvinyl alcohol-bacterial cellulose composite films. *Journal of Biobased Material and Bioenergy* 7:157–162.

Dourado, F., Leal, M., Martins, D., Fontão, A., Cristina Rodrigues, A., and Gama, M. 2016. Bacterial nanocellulose: From biotechnology to bioeconomy. *Celluloses as food ingredients/additives: Is there a room for BNC?*, eds. M. Gama, F. Dourado, and S. Bielecki, 123–133. Amsterdam: Elsevier.

Fernandes, P. 2010. Enzymes in food processing: A condensed overview on strategies for better biocatalysts. *Enzyme Research* 1–19. https://doi.org/10.4061/2010/862537.

Fontana, J.D., Koop, H.S., Tiboni, M., Grzybowski, A., Pereira, A., and Kruger, C.D., et al. 2017. New insights on bacterial cellulose. *Food biosynthesis*, ed. A.M. Grumezescu and A.M. Holban, 213–249. Cambridge: Academic Press.

Frone, A.N., Panaitescu, D.M., Nicolae, C.A., Gabor, A.R., Trusca, R., Casarica, A., Stanescu, P.O., Baciu, D.D., and Salageanu, A. 2020. Bacterial cellulose sponges obtained with green cross-linkers for tissue engineering. *Materials Science and Engineering: C* 110:110740.

Gedarawatte, S.T.G., Ravensdale, J.T., Johns, M.L., Azizi, A., Al-Salami, H., Dykes, G.A., and Coorey, R. 2020. Effectiveness of bacterial cellulose in controlling purge accumulation and improving physicochemical, microbiological, and sensorial properties of vacuum-packaged beef. *Journal of Food Science* 85:2153–2163.

Granger, C., Leger, A., Barey, P., Langendorff, V., and Cansell, M. 2005. Influence of formulation on the structural networks in ice cream. *International Dairy Journal* 15:255–262.

Guo, Y., Zhang, X., Hao, W., Xie, Y., Chen, L., Li, Z., Zhu, B., and Feng, X. 2018. Nano-bacterial cellulose/soy protein isolate complex gel as fat substitutes in ice cream model. *Carbohydrate Polymers* 198:620–630.

Halib, N., Ahmad, I., Grassi, M., and Grassi, G. 2019. The remarkable three-dimensional network structure of bacterial cellulose for tissue engineering applications. *International Journal of Pharmaceutics* 566:631–640.

Hibbert, H., and Barsha, J. 1931. Studies on reactions relating to carbohydrates and polysaccharides: xxxix. Structure of the cellulose synthesized by the action of *Acetobacter xylinus* on glucose. *Canadian Journal of Research* 5:580–591.

Hu, W., Chen, S., Zhou, B., Liu, L., Ding, B., and Wang, H. 2011. Highly stable and sensitive humidity sensors based on quartz crystal microbalance coated with bacterial cellulose membrane. *Sensors and Actuators. B* 159:301–306.

Hu, Z., Ballinger, S., Pelton, R., and Cranston, E.D. 2015. Surfactant-enhanced cellulose nanocrystal pickering emulsions. *Journal of Colloid and Interface Science* 439:139–148.

Huang, Y., Zhu, C., Yang, J., Nie, Y., Chen, C., and Sun, D. 2014. Recent advances in bacterial cellulose. *Cellulose* 21:1–30.

Iguchi, M., Yamanaka, S., and Budhiono, A. 2000. Bacterial cellulose: A masterpiece of nature's arts. *Journal of Material Science* 35:261–270.

Inoue, B.S., Streit, S., Schneider, A.L.S., and Meier, M.M. 2020. Bioactive bacterial cellulose membrane with prolonged release of chlorhexidine for dental medical application. *International Journalof Biological Macromolecules* 148:1098–1108. http://dx.doi.org/10.1016/j.ijbiomac.2020.01.036.

Jayani, T., Sanjeev, B., Marimuthu, S., and Uthandi, S. 2020. Bacterial Cellulose Nano Fiber (BCNF) as carrier support for the immobilization of probiotic, *Lactobacillus acidophilus* 016. *Carbohydrate Polymers* 250:116965. http://dx.doi.org/10.1016/j.carbpol.2020.116965. PMid:33049863.

Jebel, F.S., and Almasi, H. 2016. Morphological, physical, antimicrobial and release properties of ZnO nanoparticles-loaded bacterial cellulose films. *Carbohydrate Polymer* 149:8–19.

Jipa, I.M., Stoica-Guzun, A., and Stroescu, M. 2012. Controlled release of sorbic acid from bacterial cellulose based mono and multilayer antimicrobial films. *LWT Food Science and Technology* 47:400–406.

Jonas, R., and Farah, L.F. 1998. Production and application of microbial cellulose. *Polymer Degradation and Stability* 59:101–106. https://doi.org/10.1016/S0141-3910(97)00197-3

Jozala, A.F., Lencastre-Novaes, L.C., Lopes, A.M., Santos-Ebinuma, V.C., Mazzola, P.G., and Pessoa, Jr., A., et al. 2016. Bacterial nanocellulose production and application: A 10-year overview. *Applied Microbiology and Biotechnology* 100:2063–2072. doi: 10.1007/s00253-015-7243-4.

Juzlova, P., Martinkova, L., and Kren, V. 1996. Secondary metabolites of the fungus Monascus: A review. *Journal of Industrial Microbiology* 16:163–170.

Karahan, A.G., Kart, A., Akoğlu, A., and Çakmakçi, M.L. 2011. Physicochemical properties of low-fat soft cheese Turkish Beyaz made with bacterial cellulose as fat mimetic. *International Journal of Dairy Technology* 64:502–508. http://dx.doi.org/10.1111/j.1471-0307.2011.00718.x.

Keshk, S.M.A.S. 2014. Bacterial cellulose production and its industrial applications. *Journal of Bioprocessing and Biotechniques* 4:1–10.

Khan, S., Ul-Islam, M., Khattak, W.A., Ullah, M.W., and Park, J.K. 2015. Bacterial cellulosetitanium dioxide nanocomposites: Nanostructural characteristics, antibacterial mechanism, and biocompatibility. *Cellulose* 22:565–579.

Kim, D., and Ku, S. 2018. Beneficial effects of Monascus sp. KCCM 10093 pigments and derivatives: A mini review. *Molecules* 23:98. https://doi.org/10.3390/molecules23010098.

Kirdponpattara, S., and Phisalaphong, M. 2013. Bacterial cellulose: Alginate composite sponge as a yeast cell carrier for ethanol production. *Biochemical Engineering Journal* 77:103–109.

Klemm, D., Cranston, E.D., Fischer, D., Gama, M., Kedzior, S.A., Kralisch, D., Kramer, F., and Kondo, et al. 2018. Nanocellulose as a natural source for groundbreaking applications in materials science: Today's state. *Material Today* 21:720–748. https://doi.org/10.1016/j.mattod.2018.02.001.

Koutinas, A.A., Sypsas, V., Kandylis, P., Michelis, A., Bekatorou, A., Kourkoutas, Y., and Marchant, R. 2012. Nano-tubular cellulose for bioprocess technology development. *PLOS One* 7:343–350. https://doi.org/10.1371/journal.pone.0034350.

Kumar, Y. 2021. Development of low-fat/reduced-fat processed meat products using fat replacers and analogues. *Food Reviews International* 37(3):296–312. http://dx.doi.org/10.1080/87559129.2019.1704001.

Kumar, Y., Kairam, N., Ahmad, T., and Yadav, D.N. 2016. Physico chemical, microstructural and sensory characteristics of low-fat meat emulsion containing aloe gel as potential fat replacer. *International Journal of Food Science and Technology* 51:309–316. doi: 10.1111/ijfs.12957.

Lin, K.W., and Lin, H.Y. 2004. Quality characteristics of chinese-style meatball containing bacterial cellulose (nata). *Journal of Food Science* 69:107–111. http://dx.doi.org/10.1111/j.1365-2621.2004.tb13378.x.

Lin, S., Chen, L.C., and Chen, H.H. 2011. Physical characteristics of surimi and bacterial cellulose composite gel. *Journal of Food Process Engineering* 34:1363–1379. http://dx.doi.org/10.1111/j.1745-4530.2009.00533.x.

Lin, W.C., Lien, C.C., Yeh, H.J., Yu, C.M., and Hsu, S.H. 2013b. Bacterial cellulose and bacterial cellulose-chitosan membranes for wound dressing applications. *Carbohydrate Polymer* 94:603–611. doi: 10.1016/j.carbpol.2013.01.076.

Luo, H., Ao, H., Li, G., Li, W., Xiong, G., Zhu, Y., and Wan, Y. 2017. Bacterial cellulose/graphene oxide nanocomposite as a novel drug delivery system. *Current Applied Physics* 17:249–254. http://dx.doi.org/10.1016/j.cap.2016.12.001.

Luo, H., Feng, F., Yao, F., Zhu, Y., Yang, Z., and Wan, Y. 2020. Improved removal of toxic metal ions by incorporating graphene oxide into bacterial cellulose. *Journal of Nanoscience and Nanotechnology* 20:719–730. http://dx.doi.org/10.1166/jnn.2020.16902. PMid:31383067.

Lv, P., Feng, Q., Wang, Q., Li, G., Li, D., and Wei, Q. 2016. Biosynthesis of bacterial cellulose/carboxylic multi-walled carbon nanotubes for enzymatic biofuel cell application. *Materials* 9:183.

Ma, X., Chen, Y., Huang, J., Lv, P., Hussain, T., and Wei, Q. 2020. In situ formed active and intelligent bacterial cellulose/cotton fiber composite containing curcumin. *Cellulose* 27:9371–9382. http://dx.doi.org/10.1007/s10570-020-03413-1.

Malheiros, P.S., Jozala, A.F., Pessoa, Jr., A., Vila, M.M.D.C., and Balcão, V.M. 2018. Immobilization of antimicrobial peptides from Lactobacillus sakei subsp. sakei 2a in bacterial cellulose: Structural and functional stabilization. *Food Packaging and Shelf Life* 17:25–29. doi: 10.1016/j.fpsl.2018.05.001.

Maneerung, T., Tokura, S., and Rujiravanit, R. 2008. Impregnation of silver nanoparticles into bacterial cellulose for antimicrobial wound dressing. *Carbohydrate Polymer* 72:43–51.

Marchetti, L., and Andrés, S.C. 2021. Use of nanocellulose in meat products. *Current Opinion in Food Science* 38:96–101. http://dx.doi.org/10.1016/j.cofs.2020.11.003.

Marchetti, L., Muzzio, B., Cerrutti, P., Andrés, S.C., and Califano, A.N. 2017. Impact of bacterial nanocellulose on the rheological and textural characteristics of low-lipid meat emulsions. *Nanotechnology applications in food*, ed. A.E. Oprea and A.M. Grumezescu, 345–361. Amsterdam: Elsevier. http://dx.doi.org/10.1016/B978-0-12-811942-6.00017-0.

Matthysse, A.G., Thomas, D.L., and White, A.R. 1995. Mechanism of cellulose synthesis in Agrobacterium tumefaciens. *Journal of Bacteriology* 177:1076–1081.

Meftahi, A., Khajavi, R., Rashidi, A., Sattari, M., Yazdanshenas, M.E., and Torabi, M. 2009. The effects of cotton gauze coating with microbial cellulose. *Cellulose* 17:199–204. doi: 10.1007/s10570-009-9377-y.

Mesomya, W., Cuptapun, Y., Hengsawadi, D., Tangkanakul, P., Jittanoonta, P., and Pakpeankitvatana, R. 2002. Serum lipid-lowering in rats fed with high dietary fiber from cereal and nata de coco. *Witthayasan Kasetsart (AGRIS)* 36:187–192.

Mikkelsen, D., Flanagan, B.M., Dykes, G.A., and Gidley, M.J. 2009. Influence of different carbon sources on bacterial cellulose production by *Gluconacetobacter xylinus strain ATCC 53524. Journal of Applied of Microbiology* 107:576–583. https://doi.org/10.1111/j.1365-2672.2009.04226.x.

Mohite, B.V., and Patil, S.V. 2014. A novel biomaterial: Bacterial cellulose and its new era applications. *Biotechnology and Applied Biochemistry* 61:101–110. doi: 10.1002/bab.1148.

Montealegre, C.M., Dionisio, E.R., Sumera, L.V., Adolacion, J.R., and De Leon, R.L. 2012. A comparison between the performance of S. cerevisiae cells immobilized in nata de coco biocellulose and calcium alginate during continuous bioethanol production. *International Journal of Chemical Engineering and Application* 3:237.

Moradi, M., Tajik, H., Almasi, H., Forough, M., and Ezati, P. 2019. A novel pH-sensing indicator based on bacterial cellulose nanofibers and black carrot anthocyanins for monitoring fish freshness. *Carbohydrate Polymers* 222:115030. http://dx.doi.org/10.1016/j.carbpol.2019.115030. PMid:31320095.

Muller, A., Ni, Z., Hessler, N., Wesarg, F., Muller, F.A., Kralisch, D., and Fischer, D. 2013. The biopolymer bacterial nanocellulose as drug delivery system: Investigation of drug loading and release using the model protein albumin. *Journal of Pharmaceutical Sciences* 102:579–592.

Ng, C., Sheu, F., Wang, C., and Shyu, Y. 2004. Fermentation of Monascus purpureus on agri-by-products to make colorful and functional bacterial cellulose (NATA). *Food and Fertilizer Technology Center* 1–11. www.fftc.org.tw/htmlarea_file/library/20110716164214/tb168.pdf

Ng, C., and Shyu, Y. 2004. Development and production of cholesterol-lowering Monascus-Nata complex. *World Journal of Microbiology and Biotechnology* 20:875–879. https://doi.org/10.1007/s11274-004-0873-9.

Nguyen, D.N., Ton, N.M.N., and Le, V.V.M. 2009. Optimization of *Saccharomyces cerevisiae* immobilization in bacterial cellulose by 'adsorption-incubation' method. *International Food Research Journal* 16:59–64.

Nguyen, V.T., Gidley, M.J., and Dykes, G.A. 2008. Potential of a nisin-containing bacterial cellulose film to inhibit Listeria monocytogenes on processed meats. *Food Microbiology* 25:471–478.

Nunez, D., Cáceres, R., Ide, W., Varaprasad, K., and Oyarzún, P. 2020. An ecofriendly nanocomposite of bacterial cellulose and hydroxyapatite efficiently removes lead from water. *International Journal of Biological Macromolecules* 165:2711–2720. http://dx.doi.org/10.1016/j.ijbiomac.2020.10.055. Pmid:33069824.

Ochaikul, D., Chotirittikrai, K., Chantra, J., and Wutigornsombatkul, S. 2006. Studies on fermentation of Monascus purpureus TISTR 3090 with bacterial cellulose from *Acetobacter xylinum TISTR 967. Kmitl Science and Technology Journal* 6:13–17.

Okiyama, A., Motoki, M., and Yamanaka, S. 1992. Bacterial cellulose. II: Processing of the gelatinous cellulose for food materials. *Food Hydrocolloid* 6:479–487.

Okiyama, A., Motoki, M., and Yamanaka, S. 1993. Bacterial cellulose. IV: Application to processed foods. *Food Hydrocolloid* 6:503–511.

Oliveira, A.A.N., Mesquita, E.F.M., and Furtado, A.A.L. 2021. Use of bacterial cellulose as a fat replacer in emulsified meat products: Review. *Food Science and Technology, Campinas* 42:1–9.

Otoni, C.G., Avena-Bustillos, R.J., Azeredo, H.M.C., Lorevice, M.V., Moura, M.R., and Mattoso, L.H.C., et al. 2017. Recent advances on edible films based on fruits and vegetables: A review. *Comprehensive Reviews in Food Science and Food Safety* 16:1151–1169. doi: 10.1111/1541-4337.12281.

Panesar, P.S., Chavan, Y.V., Bera, M.B., Chand, O., and Kumar, H. 2009. Evaluation of *Acetobacter* strain for the production of microbial cellulose. *Asian Journal of Chemistry* 21:99–102.

Panesar, P.S., Chavan, Y.V., Chopra, H.K., and Kennedy, J.F. 2012. Production of microbial cellulose: Response surface methodology approach. *Carbohydrate Polymers* 87:930–934.

Paximada, P., Koutinas, A.A., Scholten, E., and Mandala, I.G. 2016a. Effect of bacterial cellulose addition on physical properties of WPI emulsions: Comparison with common thickeners. *Food Hydrocolloids* 54:245–254. http://dx.doi.org/10.1016/j.foodhyd.2015.10.014.

Paximada, P., Tsouko, E., Kopsahelis, N., Koutinas, A.A., and Mandala, I. 2016b. Bacterial cellulose as stabilizer of o/w emulsions. *Food Hydrocolloids* 53:225–232. doi: 10.1016/j.foodhyd.2014.12.003.

Pertile, R.A.N., Andrade, F.K., Alves, Jr., C., and Gama, M. 2010. Surface modification of bacterial cellulose by nitrogen-containing plasma for improved interaction with cells. *Carbohydrate Polymers* 82:692–698. doi: 10.1016/j.carbpol.2010.05.037.

Petersen, N., and Gatenholm, P. 2011. Bacterial cellulose-based materials and medical devices: Current state and perspectives. *Applied Microbiology and Biotechnology* 91:1277–1285. doi: 10.1007/s00253-011-3432-y.

Petrut, R.F., Danthine, S., and Blecker, C. 2016. Assessment of partial coalescence in whippable oil-in-water food emulsions. *Advances in Colloid and Interface Science* 229:25–33. doi: 10.1016/j.cis.2015.12.004.

Pogorelova, N., Rogachev, E., Digel, I., Chernigova, S., and Nardin, D. 2020. Bacterial cellulose nanocomposites: Morphology and mechanical properties. *Materials* 13:2849.

Purwadaria, T., Gunawan, L., and Gunwan, A.W. 2010. The production of nata colored by Monascus purpureus J1 pigments as functional foods. *Microbiology Indonesia* 4:6–10. doi: 10.5454/mi.4.1.2.

Razavi, M.S., Golmohammadi, A., Nematollahzadeh, A., Fiori, F., Rovera, C., and Farris, S. 2020. Preparation of cinnamon essential oil emulsion by bacterial cellulose nanocrystals and fish gelatin. *Food Hydrocolloids* 109:106111. http://dx.doi.org/10.1016/j.foodhyd.2020.106111.

Reis, D.T., Santos, P., Anna, K., Scheidt, G.N., and Pereira, D.H. 2019. Plant and bacterial cellulose: Production, chemical structure, derivatives and applications. *Orbital-the Electronic Journal of Chemistry* 11:321–329. doi: http://dx.doi.org/10.17807/orbital.v11i5.1349.

Rios, R.V., Garzón, R., Lannes, S.C.S., and Rosell, C.M. 2018. Use of succinyl chitosan as fat replacer on cake formulations. *LWT-Food Science and Technology* 96:2660–2265. doi: 10.1016/j.lwt.2018.05.041.

Ross, P., Mayer, R., and Benziman, M. 1991. Cellulose biosynthesis and function in bacteria. *Microbiology Review* 55:35–58.

San-Ei Gen F.F.I., Inc. 2020. www.saneigenffi.co.jp/closeup/san.html.

Shaffner, M., Ruhs, P.A., Coulter, F., Kilcher, S., and Studart, A.R. 2017. 3D printing of bacteria into functional complex materials. *Science Advances* 3:eaao6804. doi: 10.1126/sciadv.aao6804.

Shafipour, Y.A., Moradi, M., Tajik, H., and Molaei, R. 2020. Design and preparation of antimicrobial meat wrapping nanopaper with bacterial cellulose and postbiotics of lactic acid bacteria. *International Journal of Food Microbiology* 321:108561. http://dx.doi.org/10.1016/j.ijfoodmicro.2020.108561. PMid:32078868.

Sheu, F., Wang, C.L., and Shyu, Y.T. 2000. Fermentation of *Monascus purpureus* on bacterial cellulose-nata and the color stability of Monascus-nata complex. *Journal of Food Science* 65:342–345. doi: 10.1111/j.1365-2621.2000.tb16004.x.

Shi, Z., Zhang, Y., Phillips, G.O., and Yang, G. 2014. Utilization of bacterial cellulose in food. *Food Hydrocolloids* 35:539–545. doi: 10.1016/j.foodhyd.2013.07.012.

Shoukat, A., Wahid, F., Khan, T., Siddique, M., Nasreen, S., Yang, G., Ullah, M.W., and Khan, R. 2019. Titanium oxide bacterial cellulose bioadsorbent for the removal of lead ions from aqueous solution. *International Journal of Biological Macromolecules* 129:965–971. http://dx.doi.org/10.1016/j.ijbiomac.2019.02.032. PMid:30738165.

Singh, O., Panesar, P.S., and Chopra, H.K. 2017. Response surface optimization for cellulose production from agro industrial waste by using new bacterial isolate *Gluconacetobacter xylinus C18. Food Science and Biotechnology* 26:1019–1028.

Skvortsova, Z.N., Gromovykh, T.I., Grachev, V., and Traskin, Y. 2019. Physicochemical mechanics of bacterial cellulose. *Colloid Journal* 81:366–376.

Son, H.J., Heo, M.S., Kim, Y.G., and Lee, S.J. 2001. Optimization of fermentation conditions for the production of bacterial cellulose by a newly isolated *Acetobacter sp. A9* in shaking cultures. *Biotechnology and Applied Biochemistry* 33:1–5.

Ström, G., Ohgren, C., and Ankerfors, M. 2013. Nanocellulose as an additive in foodstuff. Innventia AB Report No. 403. Stockholm, Sweden, 1–25.

Sulaeva, I., Henniges, U., Rosenau, T., and Potthast, A. 2015. Bacterial cellulose as a material for wound treatment: Properties and modifications: A review. *Biotechnology Advances* 33:1547–1571. doi: 10.1016/j.biotechadv.2015.07.009.

Swazey, J.M., and Madison, N. 2010. Methods of improve the compatibility and efficiency of powdered versions of microbirous cellulose. *US Patent No 2011/0,059,883 A1.* Washington, DC: U.S. Patent and Trademark Office.

Tam, T.T.M., and Huong, N.T. 2014. Optimization of *Corynebacterium glutamicum* immobilization process on bacterial cellulose carrier and its application for lysine fermentation. *IOSR Journal of Engineering* 4:33–38.

Tanaka, M., Murakami, S., Shinke, R., and Aoki, K. 2000. Genetic characteristics of cellulose-forming acetic acid bacteria identified phenotypically as *Gluconacetobacter xylinus. Bioscience, Biotechnology, and Biochemistry* 64:757–760.

Tekin, E., Sahin, S., and Sumnu, G. 2017. Physicochemical, rheological, and sensory properties of low-fat ice cream designed by double emulsions. *European Journal of Lipid Science and Technology* 1–32. https://doi.org/10.1002/ejlt.201600505.

Tomé, L.C., Brandao, L., Mendes, A.M., Silvestre, A.J., Neto, C.P., Gandini, A., Freire, C.S., and Marrucho, I.M. 2010. Preparation and characterization of bacterial cellulose membranes with tailored surface and barrier properties. *Cellulose* 17:1203–121.

Ton, N.M.N., and Le, V.V.M. 2011. Application of immobilized yeast in bacterial cellulose to the repeated batch fermentation in winemaking. *International Food Research Journal* 18:983–987.

Trevino-Garza, M.Z., Guerrero-Medina, A.S., González-Sánchez, R.A., García-Gómez, C., Guzmán-Velasco, A., Báez-González, J.G., and Márquez-Reyes, J.M. 2020. Production of microbial cellulose films from green tea (Camellia sinensis) kombucha with various carbon sources. *Coatings* 10(11):1132.

Ul-Islam, M., Ullah, M.W., Khan, S., Kamal, T., Ul-Islam, S., Shah, N., and Park, J.K. 2016b. Recent advancement in cellulose based nanocomposite for addressing environmental challenges. *Recent Patents on Nanotechnology* 10:169–180.

Ul-Islam, M., Ullah, M.W., Khan, S., Shah, N., and Park, J.K. 2016a. Strategies for cost-effective and enhanced production of bacterial cellulose. *International Journal of Biological Macromolecules* 93:789–804.

Ullah, H., Helder, S.A., and Khan, T. 2016. Applications of bacterial cellulose in food, cosmetics and drug delivery. *Cellulose* 23:2291–2314. doi: 10.1007/s10570-016-0986-y.

Urbina, L., Guaresti, O., Requies, J., Gabilondo, N., Eceiza, A., Corcuera, M.A., and Retegi, A. 2018. Design of reusable novel membranes based on bacterial cellulose and chitosan for the filtration of copper in wastewaters. *Carbohydrate Polymers* 193:362–372. http://dx.doi.org/10.1016/j.carbpol.2018.04.007. Pmid:29773392.

Varela, P., Pintor, A., and Fiszman, S. 2014. How hydrocolloids affect the temporal oral perception of ice cream. *Food Hydrocolloids* 36:220–228. doi: 10.1016/j.foodhyd.2013.10.005.

Wang, S.S., Han, Y.H., Chen, J.L., Zhang, D.C., Shi, X.X., Ye, Y.X., Chen, D.L., and Li, M. 2018. Insights into bacterial cellulose biosynthesis from different carbon sources and the associated biochemical transformation pathways in komagataeibacter sp. W1. *Polymers* 10:963.

Wang, W., Li, H.Y., Zhang, D.W., Jiang, J., Cui, Y.R., Qiu, S., and Zhang, X.X. 2010. Fabrication of bienzymatic glucose biosensor based on novel gold nanoparticles-bacteria cellulose nanofibers nanocomposite. *Electroanalysis* 22:2543–2550.

Wang, X., Tang, J., Huang, J., and Hui, M. 2020. Production and characterization of bacterial cellulose membranes with hyaluronic acid and silk sericin. *Colloids and Surfaces B Biointerfaces* 195:111273. http://dx.doi.org/10.1016/j.colsurfb.2020.111273. PMid:32721822.

Weyell, P., Beekmann, U., Küpper, C., Dederichs, M., Thamm, J., Fischer, D., and Kralisch, D. 2019. Tailor-made material characteristics of bacterial cellulose for drug delivery applications in dentistry. *Carbohydrate Polymers* 207:1–10. http://dx.doi.org/10.1016/j.carbpol.2018.11.061. PMid:30599988.

Wonganu, B., and Kongruang, S. 2010. Red bacterial cellulose production by fermentation of Monascus purpureus. Coference proceedings, chemistry and chemical engineering (ICCCE), international conference, IEEE, Kyoto, Japan, 37–141.

Wu, S.C., and Lia, Y.K. 2008. Application of bacterial cellulose pellets in enzyme immobilization. *Journal of Molecular Catalysis B: Enzymatic* 54:103–108. https://doi.org/10.1016/j.molcatb.2007.12.021.

Wu, S.C., Lia, Y.K., and Ho, C.Y. 2013. Glucoamylase immobilization on bacterial cellulose using periodate oxidation method. *International Journal of Science and Engineering* 3:1–4. doi: 10.6159/IJSE.

Wu, S.C., Wu, S.M., and Su, F.M. 2017. Novel process for immobilizing an enzyme on a bacterial cellulose membrane through repeated absorption. *Journal of Chemical Technology and Biotechnology* 92:109–114. doi: 10.1002/jctb.4994.

Xie, Y., Niu, X., Yang, J., Fan, R., Shi, J., Ullah, N., Feng, X., and Chen, L. 2020. Active biodegradable films based on the whole potato peel incorporated with bacterial cellulose and curcumin. *International Journal of Biological Macromolecules* 150:480–491. http://dx.doi.org/10.1016/j.ijbiomac.2020.01.291. PMid:32007551.

Yan, H., Chen, X., Song, H., Li, J., Feng, Y., Shi, Z., Wang, X., and Lin, Q. 2017. Synthesis of bacterial cellulose and bacterial cellulose nanocrystals for their applications in the stabilization of olive oil pickering emulsion. *Food Hydrocolloids* 72:127–135. http://dx.doi.org/10.1016/j.foodhyd.2017.05.044.

Yu, S.Y., and Lin, K.W. 2014. Influence of bacterial cellulose (nata) on the physicochemical and sensory properties of frankfurter. *Journal of Food Science* 79:1117–1122. http://dx.doi.org/10.1111/1750-3841.12494.

Zhai, X., Lin, D., Zhao, Y., Li, W., and Yang, X. 2018a. Effects of dietary fiber supplementation on fatty acid metabolism and intestinal microbiota diversity in C57BL/6J mice fed with a high-fat diet. *Journal of Agricultural and Food Chemistry* 66:12706–12718. https://doi.org/10.1021/acs.jafc.8b05036.

Zhai, X., Lin, D., Zhao, Y., Li, W., and Yang, X. 2018b. Enhanced anti-obesity effects of bacterial cellulose combined with konjac glucomannan in high-fat diet-fed C57BL/6J mice. *Food and Function* 9:5260–5272. doi: 10.1039/C8FO01211C.

Zhai, X., Lin, D., Zhao, Y., Li, W., and Yang, X. 2018c. Bacterial cellulose relieves diphenoxylate-induced constipation in rats. *Journal of Agricultural and Food Chemistry* 66:4106–4117. https://doi.org/10.1021/acs.jafc.8b00385.

Zhang, C., Cao, J., Zhao, S., Luo, H., Yang, Z., Gama, M., Zhang, Q., Su, D., and Wan, Y. 2020a. Biocompatibility evaluation of bacterial cellulose as a scaffold material for tissue-engineered corneal stroma. *Cellulose* 27:2775–2784. http://dx.doi.org/10.1007/s10570-020-02979-0.

Zhang, T., Wang, W., Zhang, D., Zhang, X., Ma, Y., Zhou, Y., and Qi, L. 2010. Biotemplated synthesis of gold nanoparticle: Bacteria cellulose nanofiber nanocomposites and their application in biosensing. *Advanced Functional Material* 20:1152–1160.

Zhang, X., Zhou, J., Chen, J., Li, B., Li, Y., and Liu, S. 2020b. Edible foam based on pickering effect of bacterial cellulose nanofibrils and soy protein isolates featuring interfacial network stabilization. *Food Hydrocolloids* 100:105440. http://dx.doi.org/10.1016/j.foodhyd.2019.105440.

Zhong, C. 2020. Industrial-scale production and applications of bacterial cellulose. *Frontiers in Bioengineering and Biotechnology* 8:1425.

Zhu, H., Jia, S., Yang, H., Tang, W., Jia, Y., and Tan, Z. 2010. Characterization of bacteriostatic sausage casing: A composite of bacterial cellulose embedded with ε-polylysine. *Food Science and Biotechnology* 9:1479–1484. doi: 10.1007/s10068-010-0211-y.

Zhuang, S., and Wang, J. 2019. Removal of cesium ions using nickel hexacyanoferrates-loaded bacterial cellulose membrane as an effective adsorbent. *Journal of Molecular Liquids* 294:111682. http://dx.doi.org/10.1016/j.molliq.2019.111682.

11 Application of Bacterial Cellulose in the Pharmaceuticals Industry

Farheen Khan, Vartika Srivastava, Bhoomika Varshney, Vinod Pravin Sharma, and Ashok Pandey
CSIR-Indian Institute of Toxicology Research
Lucknow, Uttar Pradesh, India

11.1 INTRODUCTION

Focus on nurturing a healthy body has been a priority for a long time. Traditional medicine has been transformed by the discovery and use of synthetic drugs. However, the increasing population and changing lifestyles, as well as the associated health risks with the use of synthetic drugs, have reversed the steering of pharmaceuticals toward the use of natural resources (Ekor, 2014) Pharmaceuticals cover a broad spectrum comprising single molecules and groups of the same or different molecules in the treatment of various ailments in a manner that is least toxic to the human or animal body.

Polymers in both synthetic and natural forms have been an integral part of the development of sustainable pharmaceutics. The emergence of natural polymers. such as cellulose, natural rubber, casein, and many others, goes back to the 19th century saw. Later, the application gradually moved to synthetic polymers, due to the depletion sources of natural polymers. Yet the long-lasting health benefits of natural polymers encouraged the use of alternative sources such as bacteria, fungi, algae, marine microflora, and solid and liquid waste. The featured role of polymer in a body encircles around biocompatibility, least toxic, flexible, biodegradable, and sustainable (Ul-Isalm et al., 2012).

Cellulose is one of the prominent natural polymers utilized in controlled and delayed drug delivery and release, implants, as micro and nano capsules, gelling and stabilizing agents, and so forth. Discovered by Payen in 1938, it is the most abundant compound on earth. As a major component of the plant's cell wall, there are dramatic transformation of natural flora from utilization to exploitation, for the extraction of natural existing polymers (Song et al., 2018). The first bacterial cellulose (BC) was extracted from *Acetobacter xylinum* in 1886 by Brown (Wang et al., 2019). It is a natural nano polymeric compound with a linear chain of glucose linked with a glycosidic bond with a diameter of 20–100 nm. *What could be the best possible source of cellulose?* Among the prokaryotic and eukaryotic organisms, bacteria seem to be a better choice for the extraction of cellulose, since they have faster doubling time in minimal growth requirements. BC basically acts as a self-defense compound to protect the bacteria from ultraviolet light. Several Gram-positive and Gram-negative bacterial populations have been discovered to synthesize cellulose as part of their metabolism. The quantity of BC produced depends on the type of bacterial strain, cultivating and operational conditions, and culture medium (Moënne-Loccoz et al., 2015).

Bacterial cellulose contains high water content holding capacity (Ul-Isalm et al., 2012), high wet strength, high absorbing properties, and higher polymerization degree (Ullah et al., 2019). Moreover, it possesses higher chemical purity as compared to plant cellulose, because of the absence of lignocellulosic polysaccharides such as lignin, pectin, and hemicellulose. This provides additional benefits of simpler purification processand low energy consumption. Due to these properties (Wajid

DOI: 10.1201/9781003355434-11

et al., 2016), BC has a wide range of applications in biomedical, food packaging, environmental remediation, and paper making industries to enhance the paper quality. In relation to the biomedical applications (Waleed et al., 2013), it has been used in tissue engineering, wound healing, skin and blood vessels transplant (Zang et al., 2015), and drug delivery (Ul-Islam et al., 2012).

11.2 IMPORTANT FACTORS GOVERNING SELECTION OF OPTIMUM AND SUSTAINABLE POLYMER AS BIOMATERIAL AND BACTERIAL CELLULOSE AS EMERGENT

Any artificial implantation in the human body needs to be biocompatible, nonimmunogenic, nontoxic, absolutely biomimetic, and biodegradable. Biomaterials based on BC are excellent examples of materials with these features (Figure 11.1). The hydrophilic groups of the polysaccharide resemble the extracellular matrix and provide an optimum interconnection with neighboring tissues with superior water holding ability. Moreover, the larger surface area of the three-dimensional (3D) network provides porosity and mechanical strength to the biomaterial. Lack of cellulosic enzymes in the human body aids in enhanced the shelf-life of the BC (Petersen and Gatenholm, 2011). BC biomaterials are modified structurally based on their ex situ or in situ application (Stumpf et al., 2018), which are easily applicable due to the presence of hydroxyl groups (-OH). The mechanical or tensile strength of the biomaterial is defined by Young's modulus, a parameter that defines the elasticity limit or stretching ability, beyond which deformation of the material occurs. In this context, a recent study detailed the effect of homogenization on the improved mechanical and structural features of

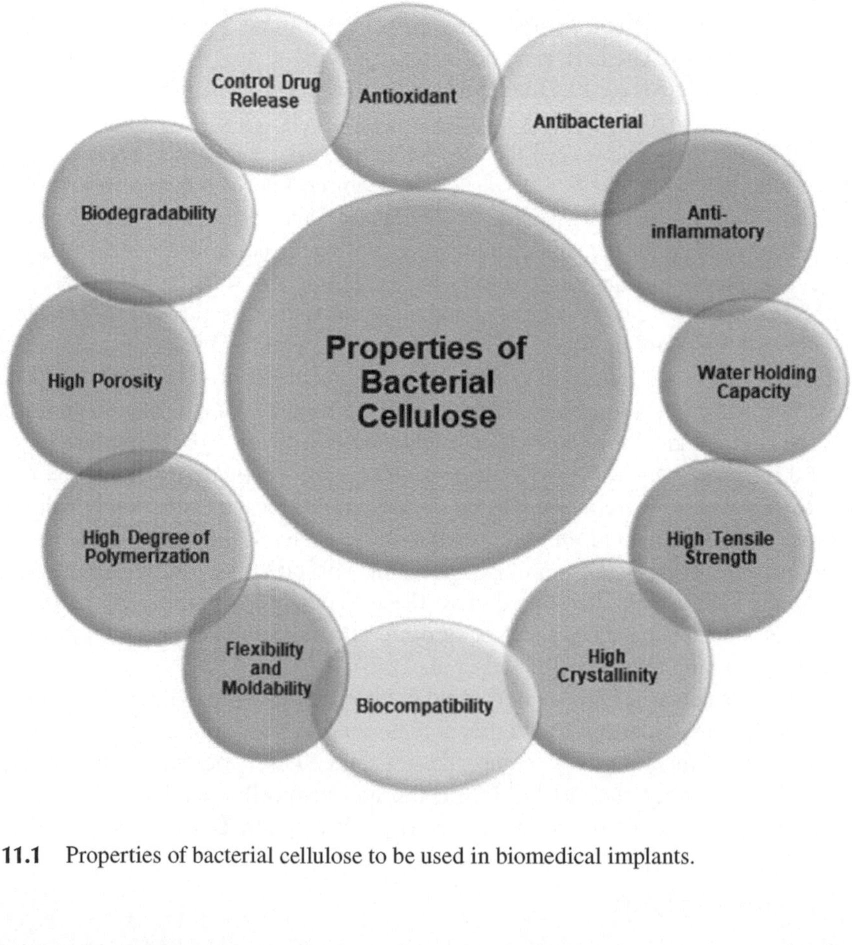

FIGURE 11.1 Properties of bacterial cellulose to be used in biomedical implants.

bacterial cellulose films. The modified BC film displayed high tensile strength, high Young's modulus, with high breaking force. Although the density of the developed material decreased, it led to better water absorption and holding capacity (Betlej et al., 2021).

11.3 THERAPEUTIC APPLICATIONS

11.3.1 Tissue Engineering

Tissue engineering, a biomedical field of growing tissues and organs in a controlled artificial environment and studying the key microphysical and chemical systems (Pang et al., 2020), came into being around 1985 (Vacanti, 2006).

11.3.1.1 Wound Healing

Wound healing mainly follows the path of homeostasis, inflammation, granulation tissue formation, followed by remodeling. When any of the above processes fails to occur, the wound becomes chronic, which usually happens in the case of ulcers. Wound healing also requires the process of elimination of exudates, avoiding infections, oxygen exchange, preventing loss of electrolytes, and maximum perspiration. The semipermeable membrane that helps in water retention in the skin is partially or completely degraded in burn patients, which leads to high evaporation rate and hypermetabolism (Lamke et al., 1977). Its 3D interwoven network allows bacterial cellulosic membrane to cover a larger surface area and provide higher moisture retention, transparency, and permeability of both liquid and gas (Portela et al., 2019). The fibrillar network also aids in the prevention of microbial infiltration and improves cell adhesion (Diegelmann and Evans, 2004). It has been revolutionary compared with the conventional gauzes and dressing materials. The first application of bacterial cellulosic dressing was BioFill® for the treatment of cuts, burns, ulcers, and dermabrasions. It is a partially dried bacterial membrane that was cost-effective, effective, and provided protection against infections, as well. Cell, a commercial product, was developed by Xylos Corporation, which developed a never-dried dressing material for the treatment of chronic ulcers. The material promoted autolytic debridement, granulation, and reduced pain. The XCell membrane participated two-way in donating and absorbing moisture for the exudates. For early relief, moist dressings are the most preferred choice with proteolytic activity that improves the concentration of growth promoting factors. As aforesaid, wound closure followed by regeneration of epidermis and dermis repair are key objectives in the case of burns. Czaja et al. (2007) found better healing effects with never-dried bacterial membrane on second-degree burn patients. It accelerated re-epithelization and tissue granulation, and reduced scarring and pain. Several efforts have been tried on limiting hemorrhages using BC-based dressings coated with hemostatic agents such as vitamin K and protamine sulfate. These biomaterials decreased lactate dehydrogenase levels, blood loss, and mortality in both diabetic and nondiabetic Sprague-Dawley rats (Karahaliloğlu et al., 2016).

Bacterial cellulose in its pure form is widely used but it has also been permeated with other polysaccharides, such as pectin, glucomannans, galactomannans, xyloglucans, and other therapeutic compounds, with the ultimate goal of accelerated wound healing. Fortification with cellulases decelerated the degradation of BC membranes. Some reports mentioned the infusion of chitosan, hyaluronic acid, superoxide dismutase, and poviargol antibiotic into BC membrane, which augmented the healing effects and reduced infection. Chitosan provided additional benefits of moisture retainment and, once degraded by endogenous enzymes, it improved angiogenesis and tissue regeneration as well. Researchers have found that Chitosan improves the tensile strength of the BC dressing, important for application on joints (Lin et al., 2013).

The applications of bacterial cellulose as temporary wound dressing material are not limited to those described here. Its potential usage for internal skin grafts has also been explored. Severely burned patients with 90% coverage require autologous skin grafts, allogenic dressings, or synthetic dressings. The synthetic skin substitutes use collagen as dressing material and is costly. BC as

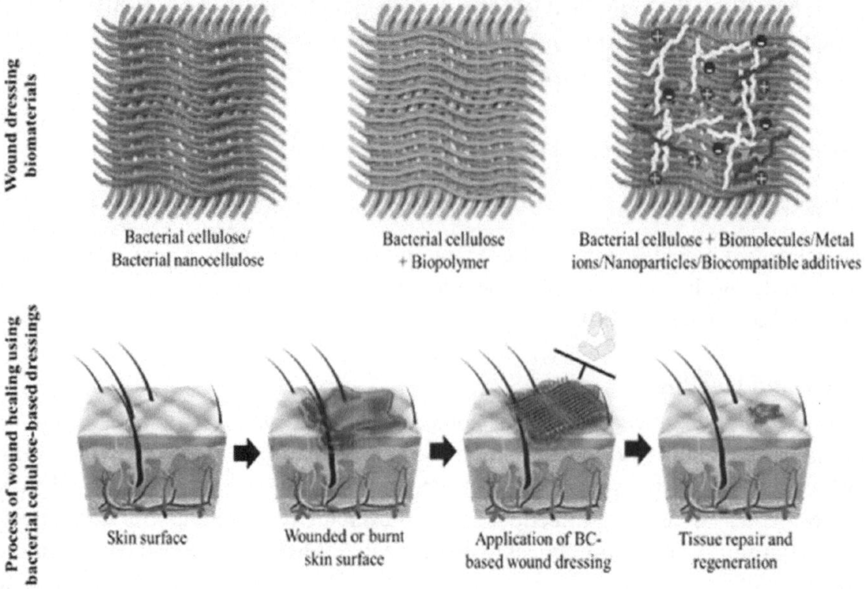

FIGURE 11.2 Role of bacterial cellulose implants in the process of wound healing.

permanent implants are impregnated well in the skin, regenerating the damaged dermis. Another important feature of BC-based wound dressings is their aid in maintaining fibroblast viability. The keratinocytes and the fibroblasts penetrate easily through the membrane, which facilitates the regeneration of the extracellular matrix. Studies have shown the sustained and delayed release of antibiotics for treating infectious wounds. The commercial applications of the BC membrane include vascular grafts, contact lenses, and drug delivery in trademarked products such as Membracell®, Suprasorb®, Bionext®, Gengiflex®, Nanoderm™, and others (Abeer et al., 2014). The Nanoderm™ is a semitransparent BC film made for the prevention of microbial attack and as an aid in tissue regeneration. It was further modified by the addition of silver nanoparticles for improved shelf-life. Overall, BC wound dressing, films, or membranes impersonate the native skin and are an effective solution for wound care, as shown in Figure 11.2.

11.3.3.2 Bone Regeneration

Bone, a living connective tissue, is a source of hematopoietic stem cells. Continuous rephasing of resorption and renewal is an exclusive feature of the bone apart from providing structure and strength to the body. However, it also undergoes abnormal tumor formations in the case of cancer or degeneration due to age or underlying disease, which hamper the mineral homeostasis, hormonal regulation, and mechanical strength (Hing, 2004). Specifically, bone tissue is composed mainly of inorganic calcium phosphates and collagen (constituting the primary bone structure), embedded in the form of a matrix, apart from other biomolecules) (Favi et al., 2013). The secondary structure of bone is in the form of large and organized collagen (Olszta et al., 2007). Traditional practices of bone regeneration include autogenic and allogenic transplantation; however, because of low acceptance rate by the host and pathogenic contamination in these traditional methods, medical practitioners have turned to the use of artificial replacements. The key requirements of artificial scaffolds or biomaterials involved in bone tissue engineering include high mechanical strength, osteoconductivity, osteoinductivity, and highly porous interconnected material. In most biomaterials developed, researchers have utilized a combination of biomaterials, such as BC membrane along with hydroxyapatite (HAp) and bone morphogenetic proteins (BMP) to treat bone defects. HAp limits the challenges of biocompatibility and osteoconduction features, while BC provides porosity and tensile strength (Lopes, 1999). BMP guide the mesenchymal cells to the site of injury where they differentiate into the osteogenic lineage, leading to the formation of new bones. Precisely, the regenerant-competent cells should have the ability of proliferation and in vitro differentiation. Its microcrystalline nature and higher water absorption capacity make BC suitable for extensive use in tissue regeneration (Barud et al., 2015).

Talking about the structural suitability of the biomimic material, there are certain key points to be considered. Although, he hybrid material provides a thorough flow of nutrients and oxygen, the brittleness and low solubility of HAp can be modified by fortification with metal ions such as Zn^{2+}, Na^{+}, Mg^{2+}, Mn^{2+}, Sr^{2+}, and CO_3^{2-}, allowing it to be more soluble in nature. The metal ions also improve the degradation and osteogenic potential. Moreover, there is an alkaline earth metal, strontium metal ion (Sr^{2+}), which is found in the bone resorption region of hard bone tissue. It has important functions in maintaining homeostasis, cell growth, and prevention of bone loss. Mostly, it is recommended to patients suffering from osteoporosis. This metal ion has been used recently with an oxidized form of BC conjugated with HAp. BC oxidation provides more active sites for Sr^{2+} binding, thereby reducing the competitive inhibition caused by Ca^{2+}, Na^{+}, and Cl^{-} ions, as well (Luz et al., 2020). This would allow the absolute release profile of Sr^{2+} for bone repair.

In the case of resorbable biomaterials, post-degradation of BC leaves a cavity that is thereby filled with collagenous substances and proteins. However, the entire process delays the process of bone repair. Irradiation of BC is another factor in causing increment in the in situ degradation process by targeting the glucose moieties. Furthermore, another defining factor is the pore size, which is governed by the type of simulated bone, cortical or cancellous. Larger pore size is always a preferred choice for better in- and outflow, keeping in mind the possibilities of cell-cell interaction. In a report by Favi et al. (2016), highly porous oxidized BC/HAp of pore diameter 300 μm decreased the cell viability from 73% to 37%.

In terms of the inclusion of growth factors as a therapeutic strategy, Dubey et al. (2021) tested the positive influence of primed bone morphogenetic protein-2 precursor (BMP-2) at a low dosage of 50 mg ml^{-1} to bacterial nanocellulose in affecting the early signs of bone matrix formation and maturation, and thereby early bone repair. In another study, fisetin, a phytoestrogen, when added to bacterial nanocellulose, enhanced osteogenic differentiation by observing the transcript expression of osteopontin and osteocalcin genes (Kheiry et al., 2018). Gelatin is a fortified biopolymer used with BC or bacterial nanocellulose. It is a derivative form of collagen and inexpensive to use in large scale. It aids in providing better mechanical strength and biodegradation properties. A higher concentration of gelatin in the biocomposite alters the calcium to phosphorus ratio, enhances surface area, and reduces crystallinity, thereby, favoring better cell adhesion followed by augmented growth and differentiation (Yang et al., 2016). Apart from this, a fibrous protein, silk fibroin, which is generally isolated from *Bombyx mori* silkworm, is also a suitable form of biopolymer that aids in improved mechanical strength, flexibility, and porosity (Jiang et al., 2018). This could also throw light on the use of aloe vera gel (*a plant source*) in obtaining a highly biocompatible, porous, and strengthened biocomposite (Foresti et al., 2017). There are few other natural and synthesized compounds that were tested for their inclusion with BC or BC-HAp biocomposite for bone regeneration (Table 11.1). The most popular is collagen, which improves HAp nucleation and aids in cell growth, but it is a rather expensive biomolecule. Graphene oxide (GO) improves the bioactivity and mechanical features and reduces the biodegradation of GO-HAp-BC nanocomposite (Ramani and Sastry, 2014). There are magnetite (Fe_3O_4) nanoparticles that aid in improving the porosity and solubility with decreased crystallinity of BC-of Fe_3O_4-HAp nanocomposite (Torgbo et al., 2019). There is also a use of osteogenic growth peptide which participates in bone repair by stimulating hematopoiesis, osteogenesis, and biomineralization (Saska et al., 2017; Reiniati et al., 2017)

TABLE 11.1
Types of Bacterial Cellulose (BC)-Based Biomaterials Depicting Bone Regeneration Potentiality

Biomaterial	Additive	Preclinical Analysis	Biological and Mechanical Properties: Increase ↑	Biological and Mechanical Properties: Decrease ↓	Reference
Oxidized BC-hydroxyapatite (Hap)	Strontium metal ion (Sr^{2+})	L-929 fibroblast cells (in vitro); biocompatibility in subcutaneous tissue from mice (in vivo)	Biodegradable and biocompatible; swelling degree of 843%; Sr^{2+} desorption in simulated body fluid (SBF) and phosphate-buffered saline (PBS) solution	Inflammatory response	Luz et al. (2020)
Oxidized BC	Strontium apatite (SrAP)		Functionally stable; biodegradable and biocompatible; formation of conjunctive tissue; swelling degree of 893%; Sr^{2+} desorption	Inflammatory response	
Oxidized BC-HAp	–	L-929 fibroblast cells (in vitro)	Biocompatible; swelling degree of 450%; connective tissue repair	Inflammatory response	Luz et al. (2020)
BC-HAp	Anti-BMP-2 (an antibody against BMP-2)	Human osteoblast MC3T3-E1 cell lines (in vitro)—Comet, and clonogenic survival assay	Nontoxic, nonmutagenic, and nongenotoxic; hydrophilicity; bone repair and biomineralization	Mechanical properties	Coelho et al. (2019)
BC-HAp	Magnetite nanoparticle (Fe_3O_4)	Mouse fibroblast L-929 and human osteoblast MC3T3-E1 cell lines (in vitro)	Biodegradable; porosity; mechanical strength; nontoxic	Swelling ability	Torgbo et al. (2019)
BC-HAp	Strontium metal ion (Sr^{2+})	–	Swelling of 961%	Sr^{2+} absorption; porosity	Luz et al. (2018)
BC	SrAP	–	Swelling degree of 1019%; slow release of Sr^{2+}	Porosity	

BC	Collagen-HAp-osteogenic peptide	Osteogenic cells of calvarial bone from 2–4 days of newborn Wistar rats, as well and CHO-K1 cell lines (in vitro)	Nontoxicity, nonmutagenicity, and nongenotoxicity; thermal stablility; osteoblastic differentiation	–	Saska et al. (2018)
BC-HAp	Silk fibroin	Human osteoblast MC3T3-E1 cell line (in vitro)	Interpenetration; tensile strength (high elastic modulus); cell attachment and proliferation; thermal stability	–	Jiang et al. (2018)
BC-HAp-BC	Glycosaminoglycan	SaOS-2, hACs and hATMSCs cell lines (in vitro) and in Wistar rats (in vivo)	Cell growth of osteoblasts and chondrocytes was observed with negative sign of immune responses; proper tissue growth in subcutaneous implantation; biocompatible; increment in bone and tissue volume	–	Kumbhar et al. (2017)
BC-Gel-HAp	–	Bone mesenchymal stem rBMSCs cell lines (in vitro)	Biodegradability; cell attachment, growth, and differentiation	–	Ran et al. (2017)
Oxidized BC-Gel-HAp	–	–	Biodegradability in SBF in 90 days	–	Yang et al. (2016)
BC-HAp	Osteogenic growth peptide	Mice for studying calvarial defects (in vivo)	Biocompatible; bone formation with no signs of inflammatory response after 60 days	–	Pigossi et al. (2015)
BC-HAp	Graphene oxide	MG63 and NIH 3T3 cell lines (in vitro)	Nontoxic; cell viability	Biodegradable; swelling ability	Ramani and Sastry (2014)
BC-Gel-HAp	–	Chondrocytes rat cells (in vitro)	Porosity; cell adhesion and proliferation	Crystallite size and 55% decrease in crystallinity	Jing et al. (2013)
BC-Gel/HAp	–	Mesenchymal stem cells (in vitro)	Cell proliferation and growth	–	Wang et al. (2011)

11.3.2 Artificial Vascular Vessel System

The cardiovascular system is the most vulnerable system of body, where slight changes in functionality can lead to fatality. The cardiovascular system is associated with blood pressure, diabetes, smoking, a stressed lifestyle, and other hormonal imbalances in the body. Moreover, the regenerative capacity of the heart is low (Doppler et al., 2017). As mentioned earlier, the ideal implant should be fully biocompatible, porous, and nonimmunogenic; in the case of the cardiovascular system it should also favor the proliferation of cardiomyocytes and avoid adverse reactions with fibroblasts, the endothelium, and perivascular cells. Also, the biomaterial should prevent myocardial dilation and scar formation. With technological improvements, the rupturing or blockage of arteries, veins, and the aorta are nowadays treated with synthetic implants. Polytetrafluoroethylene (PTFE), expanded-polytetrafluoroethylene (ePTFE), and polyethylene-terephthalates (PET) are recommended polymeric materials used in cardiovascular engineering. PTFE implants are designed for large-diameter arteries, whereas ePTFE is used for small arteries with low blood flow rate. PET forms a woven vascular graft that possesses small pore size, thereby decreasing blood leakage (Andrade et al., 2010). However, these materials are not used for <6 mm diameter arteries. The in vivo moldable nature of BC has been exploited by several researchers, who have used BC in the form of cylindrical and tubular molds in vascular grafts for the replacement of atherosclerotic coronary arteries (Klemm et al., 2001). Moreover, in pigs and rats, BC implants are used for the upkeep of carotid arteries, with ample endothelialization and no immune response (Zang et al., 2015). BC implants were inserted microsurgically in the carotid artery of rats for 1 year, which promoted the proliferation of active fibroblasts and formation of neointima as well (Schumann et al., 2009). However, modified BC biomaterials with supplementation of graphene oxide showed better angiogenesis, adhesion, proliferation rate, and biocompatibility. In a recent report, Simeoni et al. (2021) devised a BC patch laden with skeletal myoblasts and mesenchymal stem cells, which was inserted into the epicardial region of the left ventricle of Wistar rats. The cocultured cells remained viable for 14 days on the BC patch, which also provided cardioprotective soluble factors as well.

11.3.3 Food

Dietary fibers are critical for lowering health risks such as diabetes, obesity, and cardiovascular disorder. Bacterial cellulose is one of the most important dietary fiber compounds accepted by the US Food and Drug Administration (FDA) and classified as "generally recognized as safe."

The low concentration of BC avoids flavor interaction with other ingredients and enhances the stability of food. The water binding, stabilizing, thickening, and gelling properties of BC enhance the rheology of food ingredients.

When BC is added to beverages, it increases the stickiness of the product. BC retains humidity for at least one month. Added to ice cream, it helps in stabilizing the proper shape of the ice cream for at least one hour after taking it from the freezer. BChas the potential to produce low-calorie cholesteric compounds. Meatballs with 10% BC are juicy and chewy and considered the best of emulsified meat products.

11.3.4 Cosmetics

Several factors, such as ultraviolet radiation, stress, nutrition, and lifestyle have adverse effects on the skin, causing early skin aging, appearance of fine lines, and hyperchromia. Antioxidants and moisturizing active agents are used to overcome these problems. The use of modern creams is harsh to the skin as they consist of various synthetic chemicals that deprive the skin of moisture (Ramos-e-Silva et al., 2013). BC was used initially in wound healing as it provides a moist environment and

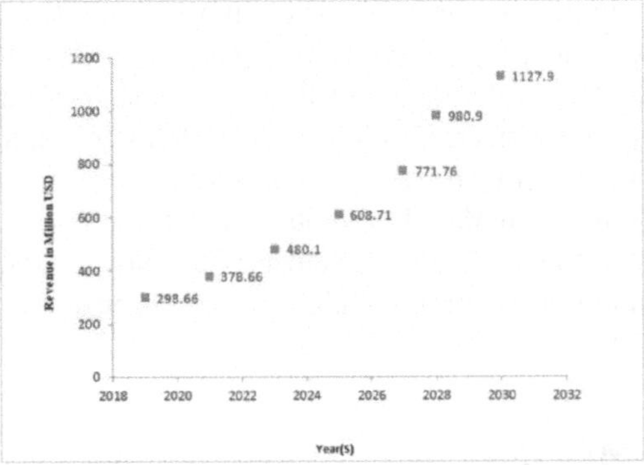

FIGURE 11.3 Projected market demand for bacterial cellulose (2018 to 2030).

is suitable for skin wound coverage. This biocompatibility is due to the fact that BC is similar to collagen and thus is effective in the treatment of wounds (Rajwade et al., 2015).

Because BC provides moisture, it can be used as an alternative for the moisturizers that we use in our day-to-day life. To validate this point Perugini and coworkers formulated a biocellulose mask for lifting, antiaging, and cell renewal. They found that skin roughness was reduced and there was an improvement in skin firmness and dermal homogeneity. Application of this mask for one month helped in cell renewal (Perugini et al., 2020). Another study led by Morais incorporated phenolic-based ionic liquids such as cholinium gallate into BC. The BC ionic liquid membranes were non-toxic and showed antioxidant and anti-inflammatory properties. They can be used to treat skin acne and breakouts (Morais et al., 2019).

BC is attracting attention in the cosmetics industry because of its promising results in skin treatment. Its prime use is in the development of face masks because it has high water holding capacity and can effectively hydrate the skin.

11.4 MARKET DEMAND FOR BACTERIAL CELLULOSE PRODUCTS

Bacterial cellulose production was valued at US $506.8 million in 2020 and it is expected to grow by 8.1% to $980.9 million by 2028. Applications for BC continue to increase in pharmaceuticals, wound healing, therapeutics, cosmetology, cardiovascular diseases, oncology, ophthalmology, urology, drug delivery systems, tissue engineering, and tissue regeneration (Petersen and Gatenholm, 2011).

At a global level, the market for BC is expected to increase by 14.80% between 2019 and 2024, from US $250 million to $570 million. The major problem with BC is its production cost which limits its commercial application (Lin et al., 2020). Several cutting-edge technologies are in the development and commercialization phases for the production of BC in different shapes and sizes, such as sheets, tubes, and fibers, for applications in different sectors. The fibrous shape of BC enhances the mechanical properties for new and advanced application. Projected market demand for BC is represented in Figure 11.3.

11.5 CONCLUSIONS

Bacterial cellulose has numerous applications in various domains, from cosmetics to food packaging and tissue engineering. It is synthesized by several nonpathogenic bacteria under controlled environmental conditions. BC is the best alternative to plant cellulose as it is biodegradable, non-toxic, and produced by preferred sustainable, renewable resources, that is agro-industrial waste.

This polymer possesses a number of properties such as biocompatibility, anti-inflammatory nature, high water retention capacity, and high porosity that make it suitable for use in diverse applications in different sectors. Currently, thin films of BC are used as face masks to rehydrate the skin. BC also is an excellent choice of material to develop scaffolds for bone regeneration and wound healing, and for use as stabilizers in various food items such as ice cream and pastries. BC-based meatballs are also available in the market. Due to its moldable nature BC is exploited to synthesize vascular implants for the treatment of atherosclerosis. Currently, studies are being conducted to further explore the application of bacterial cellulose in biosensor development, disease diagnosis, ophthalmic tissue engineering, etc.

11.6 ACKNOWLEDGMENTS

The authors are thankful to the director of CSIR-IITR for infrastructural support and access to the S&T Knowledge Resource Centre of the prestigious institute. The CSIR-IITR Communication number for this chapter is IITR/SEC/MS/2022/49.

11.7 REFERENCES

Abeer, M. M., Mohd Amin, M. C. I., & Martin, C. (2014). A review of bacterial cellulose-based drug delivery systems: Their biochemistry, current approaches and future prospects. *Journal of Pharmacy and Pharmacology*, 66(8), 1047–1061.

Andrade, F. K., Costa, R., Domingues, L., Soares, R., & Gama, M. (2010). Improving bacterial cellulose for blood vessel replacement: Functionalization with a chimeric protein containing a cellulose-binding module and an adhesion peptide. *Acta Biomaterialia*, 6(10), 4034–4041.

Barud, H. O., Barud, H. D. S., Cavicchioli, M., do Amaral, T. S., de Oliveira Junior, O. B., Santos, D. M., . . . & Ribeiro, S. J. (2015). Preparation and characterization of a bacterial cellulose/silk fibroin sponge scaffold for tissue regeneration. *Carbohydrate Polymers*, 128, 41–51.

Betlej, I., Zakaria, S., Krajewski, K. J., & Boruszewski, P. (2021). Bacterial cellulose–properties and its potential application. *Sains Malays*, 50(2), 493–505.

Coelho, F., Cavicchioli, M., Specian, S. S., Scarel-Caminaga, R. M., Penteado, L. D. A., Medeiros, A. I. D., . . . & Capote, T. S. D. O. (2019). Bacterial cellulose membrane functionalized with hydroxiapatite and anti-bone morphogenetic protein 2: A promising material for bone regeneration. *PLoS One*, 14(8), e0221286.

Czaja, W., Krystynowicz, A., Kawecki, M., Wysota, K., Sakiel, S., Wróblewski, P., . . . & Bielecki, S. (2007). Biomedical applications of microbial cellulose in burn wound recovery. In *Cellulose: Molecular and Structural Biology: Selected Articles on the Synthesis, Structure, and Applications of Cellulose* (pp. 307–321). Dordrecht: Springer Netherlands.

Diegelmann, R. F., & Evans, M. C. (2004). Wound healing: an overview of acute, fibrotic and delayed healing. *Frontiers in Bioscience*, 9(1), 283–289.

Doppler, S. A., Deutsch, M. A., Serpooshan, V., Li, G., Dzilic, E., Lange, R., . . . Wu, S. M. (2017). Mammalian heart regeneration: The race to the finish line. *Circulation Research*, 120(4), 630–632.

Dubey, S., Mishra, R., Roy, P., & Singh, R. P. (2021). 3-D macro/microporous-nanofibrous bacterial cellulose scaffolds seeded with BMP-2 preconditioned mesenchymal stem cells exhibit remarkable potential for bone tissue engineering. *International Journal of Biological Macromolecules*, 167, 934–946.

Ekor, M. (2014). The growing use of herbal medicines: Issues relating to adverse reactions and challenges in monitoring safety. *Frontiers in Pharmacology*, 4, 177. https://doi.org/10.3389/fphar.2013.00177.

Favi, C., Germani, M., & Mandolini, M. (2016). A multi-objective design approach to include material, manufacturing and assembly costs in the early design phase. *Procedia CIRP*, 52, 251–256.

Favi, P. M., Benson, R. S., Neilsen, N. R., Hammonds, R. L., Bates, C. C., Stephens, C. P., & Dhar, M. S. (2013). Cell proliferation, viability, and in vitro differentiation of equine mesenchymal stem cells seeded on bacterial cellulose hydrogel scaffolds. *Materials Science and Engineering: C*, 33(4), 1935–1944.

Foresti, M. L., Vázquez, A., & Boury, B. (2017). Applications of bacterial cellulose as precursor of carbon and composites with metal oxide, metal sulfide and metal nanoparticles: A review of recent advances. *Carbohydrate Polymers*, 157, 447–467. http://dx.doi.org/10.1016/j.carbpol.2016.09.008.

Francis, Raju, & Sakthi Kumar, D. (2016). Wiley-VCH Verlag GmbH & Co. *Biomedical Applications of Polymeric Materials and Composites*. pp. 1–389.

Gao, C., Wan, Y., Yang, C., Dai, K., Tang, T., Luo, H., & Wang, J. (2011). Preparation and characterization of bacterial cellulose sponge with hierarchical pore structure as tissue engineering scaffold. *Journal of Porous Materials*, 18, 139–145.

Hing, K. A. (2004). Bone repair in the twenty–first century: Biology, chemistry or engineering?. *Philosophical Transactions of the Royal Society of London. Series A: Mathematical, Physical and Engineering Sciences*, 362(1825), 2821–2850.

Huang, S. W., Wu, Y. F., Ahmed, T., Pan, S. C., & Cheng, C. M. (2023). Point-of-care detection devices for wound care and monitoring. *Trends in Biotechnology*. doi 10.1016/j.tibtech.2023.07.001 article in press

Jing, W., Chunxi, Y., Yizao, W., Honglin, L., Fang, H., Kerong, D., & Yuan, H. (2013). Laser patterning of bacterial cellulose hydrogel and its modification with gelatin and hydroxyapatite for bone tissue engineering. *Soft Materials*, 11(2), 173–180.

Karahaliloğlu, Z., Demirbilek, M., Ulusoy, I., Gümüşkaya, B., & Baki Denkbaş, E. (2016). Hemostatic activities of nano/microporous bilayer dressings in a femoral artery bleeding rat model. *Journal of Applied Polymer Science*, 133(28).

Kheiry, E. V., Parivar, K., Baharara, J., Bazzaz, B. S. F., & Iranbakhsh, A. (2018). The osteogenesis of bacterial cellulose scaffold loaded with fisetin. *Iranian Journal of Basic Medical Sciences*, 21(9), 965.

Klemm, D., Schumann, D., Udhardt, U., & Marsch, S. (2001). Bacterial synthesized cellulose: Artificial blood vessels for microsurgery. *Progress in Polymer Science*, 26(9), 1561–1603.

Lahori, A. H., Zhanyu, G. U. O., Zhang, Z., Ronghua, L. I., Mahar, A., Awasthi, M. K., . . . & Jiang, S. (2017). Use of biochar as an amendment for remediation of heavy metal-contaminated soils: prospects and challenges. *Pedosphere*, 27(6), 991–1014.

Lamboni, L., Li, Y., Liu, J., &re Yang, G. (2016). Silk sericin-functionalized bacterial cellulose as a potential wound-healing biomaterial. *Biomacromolecules*, 17(9), 3076–3084.

Lamke, L. O., Nilsson, G. A., & Reithner, H. L. (1977). The evaporative water loss from burns and the water-vapour permeability of grafts and artificial membranes used in the treatment of burns. *Burns*, 3(3), 159–165.

Lee, S. J., Kim, M. E., Nah, H., Seok, J. M., Jeong, M. H., Park, K., . . . & Park, S. A. (2019). Vascular endothelial growth factor immobilized on mussel-inspired three-dimensional bilayered scaffold for artificial vascular graft application: In vitro and in vivo evaluations. *Journal of Colloid and Interface Science*, 537, 333–344.

Lin, D., et al. (2020). Bacterial cellulose in food industry: Current research and future prospects. *International Journal of Biological Macromolecules*, 158, 1007–1019. https://doi.org/10.1016/j.ijbiomac.2020.04.230.

Lin, S. P., et al. (2013). Biosynthesis, production and applications of bacterial cellulose. *Cellulose*, 20(5), 2191–2219.

Lopes, M. A. F. D. S. (1999, November). Glass reinforced hydroxyapatite composites: Structural, physico-chemical characterisation and biological performance. *Biomaterials*, 20(21), 2085–2090.

Luz, E. P. C. G., Borges, M. D. F., Andrade, F. K., Rosa, M. D. F., Infantes-Molina, A., Rodríguez-Castellón, E., & Vieira, R. S. (2018). Strontium delivery systems based on bacterial cellulose and hydroxyapatite for guided bone regeneration. *Cellulose*, 25, 6661–6679.

Luz, E. P. C. G., Chaves, P. H. S., Vieira, L. D. A. P., Ribeiro, S. F., de Fátima Borges, M., Andrade, F. K., . . . & Vieira, R. S. (2020). In vitro degradability and bioactivity of oxidized bacterial cellulose-hydroxyapatite composites. *Carbohydrate Polymers*, 237, 116174.

Moënne-Loccoz, Y., et al. (2015). Environmental microbiology: Fundamentals and applications. *Microorganisms and Biotic Interactions*, 2014 Apr 29, 395–444.

Morais, E. S., Silva, N. H., Sintra, T. E., Santos, S. A., Neves, B. M., Almeida, I. F., . . . Freire, C. S. (2019). Anti-inflammatory and antioxidant nanostructured cellulose membranes loaded with phenolic-based ionic liquids for cutaneous application. *Carbohydrate Polymers*, 206, 187–197.

Olszta, M. J., Cheng, X., Jee, S. S., Kumar, R., Kim, Y. Y., Kaufman, M. J., . . . & Gower, L. B. (2007). Bone structure and formation: A new perspective. *Materials Science and Engineering: R: Reports*, 58(3–5), 77–116.

Pang, M., et al. (2020). Application of bacterial cellulose in skin and bone tissue engineering. *European Polymer Journal*, 122, 109365. https://doi.org/10.1016/j.eurpolymj.2019.109365.

Perugini, P., Bleve, M., Redondi, R., Cortinovis, F., & Colpani, A. (2020). In vivo evaluation of the effectiveness of biocellulose facial masks as active delivery systems to skin. *Journal of Cosmetic Dermatology*, 19(3), 725–735.

Petersen, N., & Gatenholm, P. (2011). Bacterial cellulose-based materials and medical devices: Current state and perspectives. *Applied Microbiology and Biotechnology*, 91, 1277–1286.

Pezzatini, S., Morbidelli, L., Solito, R., Paccagnini, E., Boanini, E., Bigi, A., & Ziche, M. (2007). Nanostructured HA crystals up-regulate FGF-2 expression and activity in microvascular endothelium promoting angiogenesis. *Bone*, 41(4), 523–534.

Pigossi, S. C., de Oliveira, G. J., Finoti, L. S., Nepomuceno, R., Spolidorio, L. C., Rossa Jr, C., . . . & Scarel-Caminaga, R. M. (2015). Bacterial cellulose-hydroxyapatite composites with osteogenic growth peptide (OGP) or pentapeptide OGP on bone regeneration in critical-size calvarial defect model. *Journal of Biomedical Materials Research Part A*, 103(10), 3397–3406.

Portela, R., Leal, C. R., Almeida, P. L., & Sobral, R. G. (2019). Bacterial cellulose: A versatile biopolymer for wound dressing applications. *Microbial Biotechnology*, 12(4), 586–610.

Rajwade, J. M., Paknikar, K. M., & Kumbhar, J. V. (2015). Applications of bacterial cellulose and its composites in biomedicine. *Applied Microbiology and Biotechnology*, 99(6), 2491–2511.

Ramani, D., & Sastry, T. P. (2014). Bacterial cellulose-reinforced hydroxyapatite functionalized graphene oxide: A potential osteoinductive composite. *Cellulose*, 21, 3585–3595.

Ramos-e-Silva, M., Celem, L. R., Ramos-e-Silva, S., & Fucci-da-Costa, A. P. (2013). Anti-aging cosmetics: Facts and controversies. *Clinics in Dermatology*, 31(6), 750–758.

Ran, J., Jiang, P., Liu, S., Sun, G., Yan, P., Shen, X., & Tong, H. (2017). Constructing multi-component organic/inorganic composite bacterial cellulose-gelatin/hydroxyapatite double-network scaffold platform for stem cell-mediated bone tissue engineering. *Materials Science and Engineering: C*, 78, 130–140.

Reiniati, I., Hrymak, A. N., & Margaritis, A. (2017). Recent developments in the production and applications of bacterial cellulose fibers and nanocrystals. *Critical Reviews in Biotechnology*, 37(4), 510–524.

Saska, S., Pigossi, S. C., Oliveira, G. J., Teixeira, L. N., Capela, M. V., Gonçalves, A., . . . & Marchetto, R. (2018). Biopolymer-based membranes associated with osteogenic growth peptide for guided bone regeneration. *Biomedical Materials*, 13(3), 035009.

Saska, S., Teixeira, L. N., de Castro Raucci, L. M. S., Scarel-Caminaga, R. M., Franchi, L. P., Dos Santos, R. A., . . . & Marchetto, R. (2017). Nanocellulose-collagen-apatite composite associated with osteogenic growth peptide for bone regeneration. *International Journal of Biological Macromolecules*, 103, 467–476.

Schumann, D. A., Wippermann, J., Klemm, D. O., Kramer, F., Koth, D., Kosmehl, H., . . . Salehi-Gelani, S. (2009). Artificial vascular implants from bacterial cellulose: Preliminary results of small arterial substitutes. *Cellulose*, 16(5), 877–885.

Simeoni, R. B., Mogharbel, B. F., Francisco, J. C., Miyague, N. I., Irioda, A. C., Souza, C. M., . . . Carvalho, K. A. (2021). Beneficial roles of cellulose patch-mediated cell therapy in myocardial infarction: A preclinical study. *Cells*, 10(2), 424.

Song, R., et al. (2018). Current development of biodegradable polymeric materials for biomedical applications. *Drug Design, Development and Therapy*, 12, 3117–3145.

Stumpf, T. R., Yang, X., Zhang, J., & Cao, X. (2018). In situ and ex situ modifications of bacterial cellulose for applications in tissue engineering. *Materials Science and Engineering: C*, 82, 372–383.

Ul-Islam, M., Khan, T., & Kon, J. (2012). Water holding and release properties of bacterial cellulose obtained by in situ and ex situ modification. *Carbohydrate Polymers*, 88(2), 596–603. http://dx.doi.org/10.1016/j.carbpol.2012.01.006.

Ullah, M. W., Manan, S., & Kiprono, S. J. (2019). Synthesis, structure, and properties of bacterial cellulose. Nanocellulose, 81–113.

Vacanti, C. A. (2006). The history of tissue engineering. *Journal of Cellular and Molecular Medicine*, 10, 569–576.

Wajid, M., et al. (2016). Structural and physico-mechanical characterization of bio-cellulose produced by a cell-free system. *Carbohydrate Polymers*, 136, 908–916. http://dx.doi.org/10.1016/j.carbpol.2015.10.010.

Waleed, A. K., Ul-islam, M., & Park, J. P. (2013). Effect of post-synthetic processing conditions on structural variations and applications of bacterial cellulose. Cellulose, 253–263.

Wang, J., Tavakoli, J., & Tang, Y. (2019, Sep 1). Bacterial cellulose production, properties and applications with different culture methods: A review. *Carbohydrate Polymers*, 219, 63–76.

Yang, X. N., Xue, D. D., Li, J. Y., Liu, M., Jia, S. R., Chu, L. Q., . . . & Zhong, C. (2016). Improvement of antimicrobial activity of graphene oxide/bacterial cellulose nanocomposites through the electrostatic modification. *Carbohydrate Polymers*, 136, 1152–1160.

Zang, S., Zhang, R., Chen, H., Lu, Y., Zhou, J., Chang, X., & Yang, G. (2015). Investigation on artificial blood vessels prepared from bacterial cellulose. *Materials Science and Engineering: C*, 46, 111–117.

12 Applications of Bacterial Cellulose in Biomedical Sector

Manoj Kumar
CSIR-Indian Institute of Integrative Medicine
Jammu, Jammu and Kashmir, India
and
Academy of Scientific and Innovative Research (AcSIR)
Ghaziabad, Uttar Pradesh, India

Ashok Pandey
CSIR-Indian Institute of Toxicology Research
Lucknow, Uttar Pradesh, India

Vinod Kumar and Saurabh Saran
CSIR-Indian Institute of Integrative Medicine
Jammu, Jammu and Kashmir, India
and
Academy of Scientific and Innovative Research (AcSIR)
Ghaziabad, Uttar Pradesh, India

12.1 INTRODUCTION

Bacterial cellulose is a class of extracellular biopolysaccharides generated by a class of bacteria and synthesized outside the cell at the cell membrane and at the cytoplasmic membrane. Bacteria produce extracellular polysaccharides in the natural environment due to logical response and selective pressures. This protects the microorganisms as they surround themselves in a highly moistened layer to protect against desiccation and predation from protozoans. Also, the existence of a gel-like biopolymer helps in diffusion.

Cellulose is basically an assemblage of cellobiose units that are linked due to removal of H_2O molecules from two β-d-glucopyranose units. Linkages of glycosidic moieties are organized connecting carbon-1 of one d-glucose unit and carbon-4 of the other. A structure called elementary fibrils, which exhibits crystalline, paracrystalline, and amorphous regions are formed from the joining of groups of cellobiose units and interconnect via hydrogen bonds (Romero et al. 2020). The first report of cellulose production from microbes from strains of *Acetobacter* was documented by Brown in 1886, who observed a gelatinous mat on the surface of a broth medium that was later found to be similar to cellulose found in the wall of plants (Liu et al. 2020). The cytoplasmic membrane is known to be the site of synthesis of bacterial cellulose by *Acetobacter xylinum;* the synthesis employ sugar nucleotide precursors and does not involve lipid intermediates in the biosynthetic pathways (Aloni et al. 1983). Organisms that produce microbial cellulose include fungi (*Saprolegnia, Dictstelium discoideum*), algae (*Vallonia*), and bacteria (*Gluconacetobacter, Acetobacter, Achromobacter, Pseudomonas, Zoogloea*). BC-producing bacteria belong to the family Acetobactereaceae, genus *Komagataeibacter* (formerly *Acetobacter* and later called *Gluconoacetobacter*), a genus known for producing high yields of cellulose. These include *G. xylinum*, *G. pasteurianus*, and *G. hansenii.*

 DOI: 10.1201/9781003355434-12

Chemically, BC and plant cellulose have a similar structure but BC has broad properties, including its high degree of polymerization, crystallinity index, and tensile strength. It is free from lignin and hemicellulose, and is nontoxic, generally regarded as safe (GRAS) by the US Food and Drug Administration, and biodegradable (Klemm et al. 2005). BC also has a high surface area per unit mass and hydrophilic attributes that result in high liquid absorbing capacity, thereby making it an interesting polymer for expansive use in the medical field (Dahman 2009). BC fibers range in size from 20–100 nm; with tensile strength of 20–300 (MPa), Young's modulus equal to 20,000 (MPa) and that of single fiber 130,000; water holding capacity greater than 95%; crystallinity within the range of 74%–96%; relative hydrophilicity of 40%–50%; purity of >99%; degree of polymerization ranging from 14,000–1,600, and porosity value of >85% (Chargot et al. 2011; Park et al. 2010).

It is assessed that the global microbial and bacterial cellulose market will reach US $771.76 million by 2027, expanding at a compound annual growth rate of 12.6% between 2019 and 2027. Diverse utilization of BC in cosmetics, cardiovascular diseases, and delivery of drugs, wound care, oncology, ophthalmology, urology, tissue engineering, and regeneration are predicted to push market broadening. The distribution of the global BC market is on the foundation of growth medium, process for production, and application. On the basis of growth medium, it is further subdivided into natural and synthetic growth medium and branched into production based on fermentation and production based on reactor. The key market players in the global bacterial cellulose market are American Process Inc., Borregaard, CelluForce, Chemfield Cellulose, Hainan Yeguo Foods Co, Innventia AB, Nippon paper, Paperlogic, and Sigachi Industries Pvt. Ltd. The key players focus on initiating and establishing ecofriendly, sustainable, and affordable products with extensive research to evolve a product with a low global impact on production and usage. Due to its high water possession property, BC has been employed for treating burns and wound infections, which has resulted in remarkable growth in the market. There has been accelerated use of BC across microbiology, biotechnology, polymer, and material science owing to its distinct mechanical, biocompatible, and biodegradable properties. It is predictable that a cost-effective and profitable growth in medical and health care industries will expand as a consequence of developments in BC-based composites and materials. Diversification of the cosmetic, food and beverage, electronics, and medical industries has made Asia Pacific the region with the largest global market share of microbial and bacterial cellulose. In the years ahead, rising demand for bacterial cellulose for manufacturing biodegradable material for the food industry is expected to spark market growth.

Bacterial cellulose has various applications in the paper, food, and electronic industries. In the recent past, BC has been used as support for immobilization of papain enzyme (Vasconcelos et al. 2020), lipase (Dikshit and Kim 2018), yeast *Yarrowia lipolytica*, *Saccharomyces cerevisiae;* (Zywicka et al. 2018), and for immobilizing antimicrobial peptides from *Lactobacillus sakei* subsp. (Malheiros et al. 2018). BC has also been used for high-efficiency particulate air filters (Liu et al. 2017) and for environmental water remediation and wastewater treatment (Alves et al. 2020). In the biomedical field BC exhibits the potential for immense application in the delivery of drugs (tetracycline hydrochloride, octenidine, diclofenac, insulin); tissue engineering; in vivo wound healing management using vaccarin, chitosan, gelatin, dextran, acrylic acid, thymol, nanoparticles; cardiovascular implants; artificial vessels; and replacement for soft tissues (Carvalho et al. 2019). Common media for the production of BC include (%w/v) glucose 2%, yeast extract 0.5%, peptone 0.5%, disodium hydrogen phosphate 0.27%, and citric acid 0.115% (Hestrin-Schramn medium); a medium having glucose 5%, yeast extract 0.5%, $(NH4)_2SO_4$ 0.5%, KH_2PO_4, 0.3%, $MgSO_4$, 0.005% (Yamanaka medium); and glucose 4%, corn steep liquor 2%, $(NH4)_2SO_4$ 0.4%, KH_2PO_4 0.2%, $MgSO_4$ 0.04% (Zhou medium). Also being used are various cheap carbon sources, molasses, sugarcane juice, and date syrup having different concentrations of additives such as ethanol, carboxy methylcellulose, agar, and alginate for maximum production (Mohammadkazemi et al. 2015).

12.2 BACTERIAL CELLULOSE–BASED HYDROGELS

Hydrogels are gels consisting of colloids formed from biopolymers and polyelectrolytes. They possess a three-dimensional polymer structure, holding an immense quantity of water, and are known to be easily recognizable and look like Jell-O®. Hydrogels act as both liquid and solid. Their physical and mechanical strength are attributed to the cross-linking present in hydrogels, which gives rise to a water insoluble character. They have hydrophilic properties due to their high water loading capacity (Ahmed 2015). Otto Wichterle is known for developing the first hydrogel in the 1950s, formed using poly (hydroxyethyl methacrylate) (PHEMA). Environmental variables, such as pH, temperature, and the contacted water ionic strength with the polymer are responsible for the quantity of H_2O retention by the network present in the hydrogel. Hydrogels are nontoxic, biocompatible, elastic in nature, and known to be high in water content, which makes it useful in the biomedical field. The degree of cross-linking, porosity, and network elasticity in the polymer is due to the presence of –OH, –COOH, $-CONH_2$, and $-SO_3H$, which are hydrophilic in nature (Hamidi et al. 2008).

The formation of hydrogen bonds due to the existence of numerous –OH groups, along with its captivating structures and increased antimicrobial properties, has made BC a candidate for preparing hydrogels. A major problem in making cellulose-based hydrogel is due to its poor, laborious dissolving in the solvent system (Edgar et al. 2001). So, there is a need to develop a successful process for the dissolution of BC. The systems described in the following sections are known to dissolve cellulose and have greatly enhanced the chance of making BC hydrogels. As BC lacks lignin and hemicellulose (which are present in plant-based cellulose) pure hydrogels can be prepared using BC.

12.2.1 *N*-Methylmorpholine-*N*-Oxide System

Dissolving high molecular weight cellulose in *N*-methylmorpholine-*N*-oxide (NMMO)/H_2O at high temperature is being employed to make a transparent solution. There is conversion of the structure of cellulose from crystalline to amorphous due to the mobility of NMMO molecules at 100°C and there is substitution of water molecules bonded to NMMO by cellulose due to the heating of the mixture at 150°C. Merely the addition of a large amount of water into the cellulose/NMMO/water solution can produce regenerated cellulose (Zhao et al. 2007).

12.2.2 LiCl/Dimethylacetamide System

To prevent notable degradation of cellulose, a nonaqueous, nonderivatizing solvent system called LiCl/dimethylacetamide (DMAc) is used. Cellulose, with molecular mass ranging from 100 to 4000 degrees of polymerization, can be readily dissolved in LiCl/DMAc. To directly construct cellulose hydrogel example of solvent system are for the pulp (7: 13 triethylammonium chloride (TEAC)/DMSO, paraformaldehyde (PF)/dimethyl sulfoxide (DMSO) for cellulose pulp (1:15), for cellulose powder (CF11) (1:9), LiCl/DMAc (Heinze et al. 2000).

12.2.3 Ionic Liquid Systems

Cellulose can be dissolved in hydrophilic salts known as ionic liquids (ILs) that show liquid behavior at temperatures lower than 100°C with properties such as nonflammability, thermal stability, and low vapor pressure. Examples of these hydrophilic ionic liquids, are 1-allyl-3-methylimidazolium chloride (AMIMCl) and 1-butyl-3-methylimidazolium chloride (BMIMCl) (Zhang et al. 2005).

12.2.4 Alkali/Urea or Thiourea Aqueous Systems

A 7 wt% NaOH/12 wt% urea solution for cellulose having molecular weight less than 114,000 and 4.6 wt% LiOH/15 wt% urea aqueous solution for cellulose having molecular weight less than

372,000 precooled up to −12°C under 2 min are being used to dissolve cellulose. Accumulation of undissolved cellulose can be seen when cellulose with molecular weight 114,000 weight% is increased above 4%, as the highest solubility of the cellulose having same molecular weight is below 4%. Another aqueous solution consisting of NaOH (9.5 wt%)/thiourea (4.5 wt%) precooled at −5°C are used to dissolve cellulose (Cai and Zhang 2006).

Based on these dissolving solvent systems BC-based hydrogels using other components to increase thermal stability and mechanical properties were reported having immense applications in biomedicine, drug delivery, and composite materials. The following points present the types of material that have been used and their application.

1. For wound healing and antimicrobial effect, silver nitrate and silver zeolite–based BC hydrogel have killing properties against *Staphylcoccus aureus* and *Escherichia coli* (Gupta et al. 2016). Zinc oxide with chitosan-based gels has shown low cytotoxicity and antibacterial activities in *E. coli* and *S. aureus* (Kim et al. 2017).
2. Collagen (Moraes et al. 2016) polyethyleneimine, lignin-based gels have wound healing and antibacterial properties, and poly (3-hydroxy-acetylthioalkanoate-co-3-hydroxyalkanoate) is used to make antistaphylococcal gels (Buceta et al. 2020).
3. Cellulose modifications to 2,3-dialdehyde cellulose impregnated with chloramphenicol, and acrylic acid and amoxicillin containing BC gels have wound healing and antimicrobial activities.
4. Curcumin and hydroxypropyl-β-cyclodextrin–loaded BC gels can be used as a wound dressing with hemocompatibility and cytocompatible properties, and plant extract–based BC gels with *Calendula officinalis*, commonly known as pot marigold, and *Jacaranda caroba* have excellent wound-healing activities (Moraes et al. 2016).
5. For producing hydrogels for drug delivery systems. Gelatin, polyacrylamide, hyaluronic acid, poly(vinyl alcohol), acrylic acid for an oral delivery system for proteins and graphene oxide are being used to synthesize thermally stable, having good increased mechanical, tensile strength, compression properties, and thermal stability. Nanostructured lipid carriers (NLCs) including doxorubicin are being used for delivery of drugs and in localized cancer treatment (Cacicedo et al. 2018).
6. Cellulose-based hydrogels have also applications in electronics as copper nanoparticles and alginate beads can be used as antimicrobial three-dimensional printing material and titanium dioxide (TiO_2) nanoparticles have applications in electronics and sensors (Gutierrez et al. 2019; Brandes et al. 2016).
7. For three-dimensional cell-culture platform hydroxyapatite and methacrylated gelatin are used in cartilage tissue engineering (Athukorala et al. 2021).

12.3 BACTERIAL CELLULOSE COMPOSITES (COLLAGEN, CHITOSAN, GELATIN, VACCARIN, AND NANOPOROUS CELLULOSE)

Gelatin derived from the extracellular matrix, collagen, along with bacterial cellulose is used to increase mechanical strength, the reversible swelling ability after repeated compression (Nakayama et al. 2004). A sponge-like composite has been synthesized with gelatin incorporated into BC with the use of a cross linker, namely, glutaraldehyde. The BC gelatin sponges synthesized have a uniform honeycomb-like assembly having high porosity of 94%–95% and swelling properties ranging from 3000%–3150%. The sponge shows antibacterial activity against *Escherichia coli, Candida albicans,* and *Staphylcoccus aureus* and can be used in wound care. Cross-linking using procyanidin has also been used to synthesize 0.25 wt.% gelatin-based BC composites and their adhesion and proliferation on NIH3T3 cells evaluated. Silver nanoparticles capped with gelatin have better stability and antibacterial activity used for the treatment of *S. aureus*-induced keratitis (Luo et al. 2019).

Chitosan, a polysaccharide acquired from the *N*-deacetylation of chitin, composed of polymeric (1→4)-linked 2-amino-2-deoxy-_-d-glucopyranose units, has been employed in synthesizing antibacterial paper using bacterial cellulose. Chitosan having intrinsic antibacterial properties, is regarded as a flexible substance with possible biomedical applications due to its biodegradability, nontoxicity, biocompatibility, and antibacterial properties. Bacterial cellulose-chitosan composites were developed by dipping BC sheet into an aqueous solution of chitosan and by impregnating BC pulp with chitosan prior to production. The synthesized nanocomposites showed antimicrobial, antioxidant, and increased physical properties (Romero et al. 2020). In another application, of the Schiff base reaction was used to graft chitosan onto BC, and nano-sized fibrils were obtained using an ultrasonic dispersion process. A 12.38% of maximum grafting amount was obtained using different concentrations of $NaIO_4$. Better acid and temperature stability were observed in suspensions having chitosan. Chitosan-based BC composites have also been used in wound care applications to significantly reduce the growth of *S. aureus* and *E. coli*. BC-chitosan composites had no cytotoxicity demonstrated by cell culture. In skin wound healing evaluated in rat models, BC-chitosan increases epithelization and regeneration in comparison to models treated with BC only, and BC-chitosan maintains proper moisture content for longer periods without dehydration.

Collagen is a natural biological polymer present in most tissues and organs, and is used extensively as a substitute for skin in wound care management. It increases cell proliferation and cell adhesion but has some drawbacks, such as less mechanical strength and higher price. BC-based collagen composites are being synthesized to compliment both polymer properties. BC sheets impregnated with collagen have improved thermal stability. Composite scaffolds having collagen on BC give favorable results, with cell attachment and growth, and can take up to 400% of water. The composites demonstrate a porous structure having well-organized interconnection when synthesized in BC:collagen ratios of 1:1, 3:1, and 5:1 (Noh et al. 2019) and showed cytocompatibility in cell adhesion studies, when used with 3T3 fibroblast cells. The composite has vastly superior cytocompatibility than pristine BC and was effective in adhesion and cell proliferation after 48 h of incubation. These BC/collagen scaffolds are suitable for infection dressing, tissue scaffolds, and for cell adhesion, proliferation and attachment.

Antibacterial BC membranes for preventing wound infection are also being synthesized by dipping BC in vaccarin solution. Vaccarin, known for endothelial cell proliferation along with neovascularization in vitro, is a glycoside flavonoid found in vaccariae semen, and can diminish endothelial oxidative stress. Wound treatment using BC-Vac accelerates regeneration and epithelization more in rat models than in wounds treated with only BC. BC-Vac is a good candidate for cell growth and it has no cytotoxicity inferred from cell viability and cell attachment (Xie et al. 2015). Another scaffold, namely microporous bacterial cellulose 300–500 μm prepared in fermentation using microspheres obtained from paraffin wax that has increased the cell seeding by enabling larger concentration of cells in pores when seeded with MC3T3-E1 osteoprogenitor cells and has shown to form denser mineral deposits in comparison to cells grown on BC only. Thus, microporous bacterial cellulose are an appealing upcoming scaffold for bone regeneration (Zaborowska et al. 2010).

12.4 BACTERIAL CELLULOSE METAL NANOCOMPOSITES AND NANOPARTICLES

A severe risk to public health worldwide is seen due to the surfacing of infectious diseases. Both Gram-positive and -negative strains are responsible for extensive public health complications. In the course of time antibiotics have been used to restrict infections arising from both community and hospital environments (Hawkey 2008). Innumerable reports have been proclaimed to ameliorate the present antimicrobial therapies owing to the arrival of antibiotic-resistant pathogens that have become an alarming health issue. It is documented that bacterial infections are over 70% resistant to antibiotics that are frequently used to eliminate the isepticemia (Allahverdiyev et al. 2011). For many thousands of years, people have used metals in controlling infection. Egyptians in 1500 BP first recorded the use of copper salts as an astringent (Lemire et al. 2013).

The antimicrobial activity of metals having diverse applications, efficacy, and antimicrobial activity has been established and practiced for generations. In the recent past, nanotechnology has extended significant prospects among numerous fields of life science and biomedicine. Pharmaceutical nanotechnology with diverse benefits has be increasingly recognized by scientists (Adibkia et al. 2007).

Metals have been used extensively along with bacterial cellulose to form metal nanoparticles and BC+ metal nanocomposites having antibacterial, wound healing, increased mechanical strength, and improved properties. Metals along with other substances that have been used by different researchers include copper oxide, gold(III) chloride solution (Chen et al. 2015), silver chloride (Hu et al. 2009), magnesium oxide (Mirtalebi et al. 2019), copper chloride (Phutanon et al. 2019), silver nitrate, silver ammonia solution (Wu et al. 2014), zinc oxide (Jebel and Almasi 2016), silver nitrate with polyaniline, silver nanopowder with chitosan (Salari et al. 2018a), silver nitrate with calcium alginate (Yang et al. 2017), and carbon quantum along with dots-titanium dioxide (Malmir et al. 2020). Zinc oxide nanoparticles formed using zinc acetate and diethylene glycol (Wahid et al. 2019), $CuCl_2$ solution for synthesizing copper nanoparticles (He et al. 2018), and zinc acetate and acetic acid to produce zinc sulfide (ZnS) nanoparticles (Zheng et al. 2014). The process that has been used for producing these composites are mainly chemical reduction using $NaBH_4$, the polyol method (Wahid et al. 2019), and the green hydrothermal approach in which BC itself serves as a reducing and stabilizing biopolymer and the BC matrix acts as a nanoreactor for metal reduction and successful impregnation onto the 3D bacterial cellulose matrix (Araujo et al. 2018; Wu et al. 2014).

BC-based metal nanocomposites have increased physical and antibacterial properties. Metal and other materials that have been used and different processes involved in synthesizing metal-based nanocomposites are represented in Table 12.1. For characterization of metal-based nanocomposites,

TABLE 12.1
Process Involved in Synthesizing Metal Nanocomposites and Their Applications

Metal Used	Process and Chemical Used	Application	Reference
Silver chloride	In situ synthesis	Antimicrobial activity against *Escherichia coli* and *Staphylococcus aureus*	Hu et al. (2009)
Copper oxide	Hydrothermal	Conversion of cellulose into lactic acid and acetic acid	Li et al. (2012)
Silver nitrate	Hydrothermal	Antibacterial application	Wu et al. (2014)
zinc sulfide (ZnS) nanoparticles from zinc acetate and acetic acid	In situ precipitation method	High photocatalytic activity (92%) and photocatalysis	Zheng et al. (2014)
Gold(III) chloride solution	Amidoxime	Increased catalytic activity, 4-nitrophenol (4-NP) reduction using $NaBH_4$	Chen et al. (2015)
Silver nitrate	In situ synthesis	Sensing of CN^- and 2-mercaptobenzothiazole (MBT) via development of probe	Pourreza et al. (2015)
Zinc oxide	ZnO dispersion	Novel food packaging materials	Jebel and Almasi (2016)

(Continued)

TABLE 12.1 (*Continued*)
Process Involved in Synthesizing Metal Nanocomposites and Their Applications

Metal Used	Process and Chemical Used	Application	Reference
Silver nitrate	$NaBH_4$ reduction	Shelf life of tomatoes	Adepu and khandelwal (2017)
Copper oxide	Hydrothermal	Antibacterial against *E. coli, Staphylcoccus aureus,* and *Salmonella*	Araujo et al. (2018)
Sodium hydroxide (NaOH) and zinc nitrate tetrahydrate	Chemical reduction	Wound care activity against *Pseudomonas aeruginosa, Citrobacter freundii, S. aureus,* and *E. coli*	Khalid et al. (2017)
Silver nitrate and calcium alginate	In situ synthesis	Biomedical application	Yang et al. (2017)
Copper nanoparticles form $CuCl_2$ solution	In situ chemical reduction method	Antibacterial activity against *E. coli* and *S. aureus*	He et al. (2018)
Silver nanopowder with chitosan	In situ synthesis	Improved mechanical properties, food packaging for shelf life	Salari et al. (2018a)
Magnesium oxide	*In situ* synthesis *Ex situ* MgO dispersion	Antibacterial against *Staphylococcus aureus* and *Escherichia coli*	Mirtalebi et al. (2019)
Copper chloride	Forced hydrolysis technique	Photocatalytic and antibacterial properties against *E. coli* and *S. aureus*	Phutanon et al. (2019)
Using zinc acetate, diethylene glycol to form zinc oxide nanoparticles	Polyol method	Degradation of methyl orange	Wahid et al. (2019)
Silver nitrate with polyaniline	Hydrothermal reduction	Soft tissue engineering	Hosseini et al. (2020)
Silver nitrate	Catecholic redox chemistry	Wound healing of third-degree burn in rats	Jiji et al. (2020)
Carbon quantum dots-Titanium dioxide	Chemical reduction	Antibacterial	Malmir et al. (2020)

techniques such as atomic force microscopy, scanning electron microscopy for surface morphology, transmission electron microscopy for particle size and distribution, Fourier transform infrared microscopy for chemical bonding, functional group, x-ray diffraction for the structure of crystalline material, thermogravimetric analysis for mechanicaland thermal properties have been used to confirm successful impregnation and uniform distribution of metal-based composite nanoparticles on bacterial cellulose matrix.

Synthesized BC-metal nanocomposites have extensive usage in the field of biomedicine acting as an antibacterial wound care material, inhibiting the growth of pathogens. These nanocomposites are effective against *Staphylcoccus aureus, Pseudomonas aeruginosa, Citrobacter freundii, Escherichia coli,* and *Salmonella* (Araujo et al. 2018; Chen et al. 2009; Khalid et al. 2017; Wu et al. 2014). Silver nitrate synthesized nanoparticles have wound healing properties in third-degree burns in rats and can be used in soft tissue engineering (Hosseini et al. 2020) and as a probe for the optical sensing of cyanide ion (CN^-) and 2-mercaptobenzothiazole (MBT). In the food industry, silver nanoparticles impregnated into BC sheets are used for food packaging, fruit storage, and increasing the shelf life of tomatoes (Adepu and Khandelwal 2017). Another that has been widely used is zinc.

Zinc oxide and zinc sulfide nanoparticles containing bacterial cellulose membranes are used for the reduction of 4-nitrophenol (4-NP) with the help of $NaBH_4$, have increased catalytic activity (Chen et al. 2015), and has application in methyl orange degradation and photocatalytic activity (Wahid et al. 2019).

12.5 DRUG DELIVERY

Researchers all over the world have used BC for drug delivery systems. Because of its biodegradability, biocompatibility, and nontoxic nature, it has been considered a favorable biopolymer for drug delivery. Silva et al. (2014) used diclofenac sodium salt (0.5%, 1%), a nonsteroidal anti-inflammatory drug, for loading onto BC using 5% glycerol as plasticizer. The transdermal membrane produced showed flexibility, increased mechanical properties with Young modulus (MPa) of 1333.34 ± 361.66, tensile strength of 49.80 ± 11.68, and elongation at break of 6.75% ± 1.35%. The results showed similar permeation from the rates to those obtained from commercial patches. Müller et al. (2013) used for protein drug delivery using serum albumin as model drug. Albumin drug loading was found to be more in native BC than freeze-dried cellulose and gel electrophoresis was used to indicate the retention of biological activity during loading and release processes. Treesuppharat et al. (2017) developed a 10% gelatin-based bacterial cellulose composite using glutaraldehyde (1%) for cross-linkage for loading methylene blue as a model drug into the composite using the swelling-diffusion method. The results showed increased thermal behavior, mechanical properties, and chemical resistance, and the swelling ratio of the developed network of hydrogel in water was estimated to be 400%–600%. For drug storage Narh et al. (2019) introduced inulin to the fructose permeates of *Acetobacter xylinum* to synthesize an empty bacterial cellulosic pocket with higher heat resistance than the normal bacterial cellulose as showed by thermal analysis.

12.6 TISSUE ENGINEERING

Serious health problems arise from tissue and bone defects caused by bruising, damage, injury, and infection. In several cases there is demand for bone regeneration to recover activity. Tissue engineering depends on transplating biofactors along with 3D porous biomaterial known as scaffold for cell attachment, support, and providing a suitable environment for the regeneration of tissues. Over the years, biopolymers have gained worldwide attention in biomedicine for their potential use in preference to synthetic polymer products. Klinthoopthamrong et al. (2020) grafted BC with poly(acrylic acid) to develop a nonresorbable guided tissue regeneration membrane by associating human osteopontin, a plant-derived recombinant to multiple carboxyl groups formed after poly(acrylic acid) treatment. The quantification of adhered osteopontin was carried out using enzyme-linked immunosorbent assay and suggested its usage to promote bone tissue regeneration. Frone et al. (2020) devised a new procedure for obtaining highly porous 3D sponge structures employing BC by using glucose, citric acid, and vanillin as cheap cross-linkers and partial replacement of OH from the surface of BC with amino groups by aminopropyltriethoxysilane. Strength to weight ratios of 150% and 120% were derived with glucose and citric acid, respectively. The noncytotoxic and noninflammatory response of the modified BC sponges in macrophages are affirmed by in vitro assays.

A BC and silk fibroin-based sponge composite for tissue regeneration was developed by Barud et al. (2015). The cytocompatibility of the composite was investigated by assessing cell metabolic activity. A 3D porous microsphere was also developed by (Zhang et al. 2020) by recombining BC and collagen, responsible for healthy joints and elasticity in the body, by malaprade and Schiff-base reactions. Reverse-phase suspension regeneration was employed to develop COL/BC/bone morphogenetic protein 2 (BMP-2) microspheres with particle size ranging from 8 to 12 microns, and the average diameter of pore was 198.5 nm. The use of collagen for bone tissue regeneration has also been reported by Saska et al. (2012) by cross-linking type I collagen using 1-ethyl-3-(3-dimethylaminopropyl)-carbodiimide on BC matrix. In another report, BC along with magnetite

(Fe^3O^4) and hydroxyapatite (HA) nanoparticles was used to develop a nanocomposite scaffold through ultrasonic irradiation for bone tissue regeneration (Torgbo and Sukyai 2019) with calcium to phosphorus (Ca/P) ratio of 1.63 (surface) and 1.56 (cross section). The nontoxic nature and biocompatibility of the scaffold was illustrated by cell culture experiment using mouse fibroblast L929 cells and osteoblast (MC3T3-E1 cell line), respectively. The use of multiwalled carbon nanotubes with bacterial cellulose for in vitro assessment of osteoblastic cells for suitable 3D scaffold showed high adhesion and proliferation in contrast to conventional culture substrates. BC-modified polyhydroxyalkanoate (PHB/BC) scaffolds for mouse calvaria defects revealed scaffolds support for 3T3-L1 preadipocyte proliferation, new bone formation induction 20 weeks after implantation and in vivo osteoblast differentiation (Codreanu et al. 2020).

12.7 WOUND HEALING

Coagulation, inflammation, cell proliferation, and epithelialization are steps involved in the healing process of wounds. Healing can extend up to several months depending on the extent of skin damage. To regain the functioning of damaged skin, polymer-based dressings have been developed to cover the wounded area, minimize the spread of infection, and increase healing. High liquid loading capacity, high mechanical strength in the moisture state, liquid permeability, and nontoxicity make BC a good candidate as an artificial skin for short-term covering of wounds. BC itself lacks antibacterial activities but due to its various properties it is used in combination with other components for wound healing agents. Biocellulose membranes containing 1.2%, 2.4%, and 3.6% (w/v) of Brazilian propolis were developed for wound healing (Barud et al. 2013). Uptake, delivery, and release of octenidine, povidone-iodine, and fluorescein isothiocyanate (FTIC)-dextran molecules (10 kDa, 70 kDa, and 2000 kDa) were tested with BC as wound dressing material. The results showed the loading of antiseptic onto BC depends inversely on molecular weight. Faster release of povidone-iodine was observed and prolonged delivery on solid matrix was observed for octenidine and they showed dose-dependent antibacterial effect against *S. aureus*. Impregnation of coffee extract having polyphenols content of 13.66 mg/g onto BC for developing wound dressing exhibits minimum inhibitory concentration, with 86% and 97% growth inhibition against *E. coli* and *S. aureus*, respectively (El-Wakil et al. 2019) and kaolin-BC nanocomposites also showed wound-healing properties in which BC acts as a binding agent for kaolin particles (Wanna et al. 2013). Wichai et al. (2019) incorporated sodium alginate, chitosan, and copper sulfate on BC to develop antibacterial wound material effective against both Gram-positive methicillin-resistant *S. aureus* (MRSA) and the Gram-negative bacteria *E. coli*. Antibiotics such as ciprofloxacin, ceftriaxone, tetracycline hydrochloride (TCH), and amoxicillin, and carbon nanomaterials such as graphene oxide along with chitosan are also used to produce antibacterial wound dressing. Organic antimicrobial compounds such as curcumin (Gupta et al. 2019) showed antibacterial activity against *S. aureus, P. aeruginosa, and Candida*. Biofill®, Bioprocess®, Gengiflex®, and XCell® are commercially available wound dressings from BC. These products show better adhesion to the wound surface, diminished infection rate, and better healing.

12.8 ARTIFICIAL BLOOD VESSELS

Bioengineers have researched alternative materials for use as artificial blood vessels to restore the function of injured tissue. Bacterial cellulose can be used as for artificial blood vessels owing to its distinct mechanical properties in addition to shape retention and tear strength, which are better than many synthetic materials. The purity, nontoxicity, biocompatibility, ultrafine fibrous structure, and higher tensile strength make BC a potential tubular scaffold material. For atherosclerotic coronaries, artificial vessels were developed from BC using cylinder glass molds (Klemm et al. 2001). For rats and pigs bacterial synthesized cellulose (BASYC®) were successfully employed for implantation and extended maintenance of carotid arteries. Reports on in vivo biocompatibility testing

TABLE 12.2
Biomedical Applications of Bacterial Cellulose–Based Material

Modification	Outcome/Application	Reference
BC-chitosan, (PEG), gelatin	Scaffold for tissue engineering, wound dressing	Kim et al. (2010)
BC-adhesion peptide	Blood vessel	Andrade et al. (2010)
BC/polyvinyl alcohol	Artificial cornea	Wang et al. (2010)
BC-silk nanofibers	Wound dressing	Park et al. (2012)
BC-chondroitin sulfate	Dental scaffold	de Olyveira et al. (2014)
BC-hydroxyapatite	Bone tissue regeneration	Duarte et al. (2015)
BC-graphene oxide	Biocompatible, antibacterial	Shao et al. (2015)
BC-chitosan	Wound dressing	Ardila et al. (2016)
BC-vaccarin	Tissue engineering	Qiu et al. (2016)
BC-poly(acrylamide)	Bone repair, oral drug delivery	Yuan et al. (2016)
BC-piezoelectric immunochip	Dengue detection	Pirich et al. (2017)
BC-carboxymethyl cellulose (CMC)	Drug delivery	Fontes et al. (2019)
BC-poly-L-lysine	Antimicrobial	Fürsatz et al. (2018)
BC/poly (hydroxyethyl methacrylate hydrogel matrix	Tissue replacement	Hobzova et al. (2018)
BC-hyaluronic acid	Artificial cornea	Jia et al. (2018)
BC-chitin	Biocompatible, surgical sutures	Wu et al. (2018)
BC-keratin	Wound healing	Azarniya et al. (2019)
BC-gelatin	Tissue regeneration	Gorgieva and Hribernik (2019)
BC-acrylic acid	Wound dressing	Mohamad et al. (2019)
BC-collagen	Scaffold for bone	Noh et al. (2019)
BC-fibrinogen	Bone tissue engineering	Osorio et al. (2019)
BC-chlorhexidine	Bioabsorbable membrane for dental use	Inoue et al. (2020)
BC-alginate	Biomedical	Wei et al. (2020)

indicate no signs of inflammation and complete endothelialization around the implants (Lee and Park 2017). In the search for new vascular graft material, BC has been explored by assessing its interactions with cells and blood. Surface modifications of BC were performed to investigate thrombogenic properties and human endothelial cells assisting in contrast to conventional graft materials. Table 12.2 and Figure 12.1 summarize some biomedical applications of BC-based material.

12.9 CONCLUSIONS

In recent times, there has been increased demand in the health and biomedical sectors for alternatives to traditional methods, challenging researchers to look for alternatives and new systems for delivering drugs, tissue engineering, and controlling infections. The use of plant-based cellulose to fulfill the immense demand has a deleterious effect on the environment with increased global warming and deforestation. Bacterial cellulose is an alternative to plant cellulose that has high purity and is known for various properties such as its high liquid loading capacity and nontoxicity. The production of bacterial cellulose on a laboratory scale is easy and efficient, utilizing carbon (glucose, fructose, sucrose) and nitrogen (yeast extract, peptone, corn steep liquor) sources with the addition of some salts. BC-based composites with increased mechanical, thermal, physical,

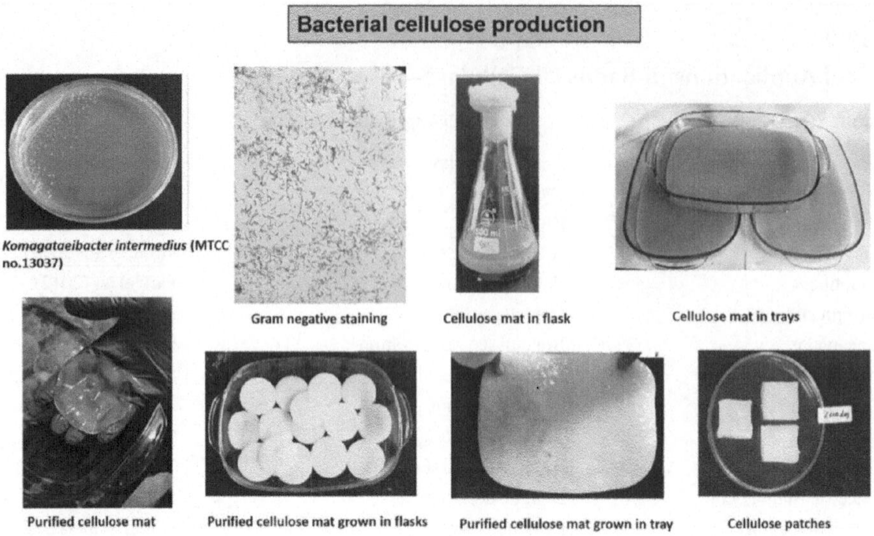

FIGURE 12.1 Biomedical applications of bacterial cellulose.

and antibacterial properties have opened new doors for the health sector. Because of its low toxicity, biodegradability, low cytotoxicity, and the biocompatibility of BC-based hydrogels, BC-based antibacterial nanocomposites, 3D porous nanospheres, and drug delivery systems have enormous applications. BC-based materials can fulfill the health sector demand with the development of eco-friendly new systems.

12.10 REFERENCES

Adepu S, Khandelwal M (2017) Broad-spectrum antimicrobial activity of bacterial cellulose silver nanocomposites with sustained release. J Mater Sci. 53:1596–1609.

Adibkia K, Omidi Y, Siahi MR, Javadzadeh AR, Jalali MB, Barar J, Maleki N, Mohammadi G, Nokhodchi A (2007) Inhibition of endotoxin-induced uveitis by methylprednisolone acetate nanosuspension in rabbits. J. Ocul. Pharmacol. Ther. 23(5):421–432.

Ahmed EM (2015) Hydrogel: Preparation, characterization, and applications: A review. J Adv Res. 6(2):105–121.

Allahverdiyev AM, AbaMor ES, Bagirova M, Rafailovich M (2011) Antimicrobial effects of TiO2 and Ag2O nanoparticles against drug-resistant bacteria and leishmanial parasites. Future Microbiol. 6:933–940.

Aloni Y, Cohen R, BenziMan M, Delmer D (1983) Solubilisation of the UDP-glucose: 1,4-beta-D-glucan 4-beta-D-glucosyltransferase (cellulose synthase) from *Acetobacter xylinum*. A comparison of regulatory properties with those of the membrane-bound form of the enzyme. J. Biol. Chem. 258:4419–4423.

Alves AA, Silva WE, Belian A, Lins LSG, Galembeck A (2020) Bacterial cellulose membranes for environmental water remediation and industrial wastewater treatment. Int. J. Environ. Sci. Technol. 17:3997–4008.

Andrade FK, Costa R, Domingues L, SoaRes R, Gama M (2010) Improving bacterial cellulose for blood vessel replacement: Functionalization with a chimeric protein containing a cellulose-binding module and an adhesion peptide. Acta Biomater. 6:4034–4041.

Araujo IMS, Silva RR, Pacheco G, Lustri WR, TercjAk A, Gutierrez J, Junior JRS, Azevedo FHC, Figueredo GS, Vega ML, Ribeiro SJL, Barud HS (2018) Hydrothermal synthesis of Bacterial Cellulose: Copper oxide nanocomposites and evaluation of their antimicrobial activity. Carbohydr. Polym. 179:341–359.

Ardila N, Medina N, Arkoun M, Heuzey MC, Ajji A, Panchal CJ (2016) Chitosan: Bacterial nanocellulose nanofibrous structures for potential wound dressing applications. Cellulose. 23:3089–3104.

Athukorala SS, Liyanage CJ, Jayasundera ACA (2021) Hydroxyapatite incorporated bacterial cellulose hydrogels as a cost-effective 3D cell culture platform. Soft Mater. 1–10.

Azarniya A, Tamjid E, Eslahi N, Simchi A (2019) Modification of bacterial cellulose/keratin nanofibrous mats by a tragacanth gum-conjugated hydrogel for wound healing. Int. J. Biol. Macromol. 134:280–289.

Barud Hda S, de Araújo Júnior AM, Saska S, Mestieri LB, Campos JA, de Freitas RM, Ferreira NU, Nascimento AP, Miguel FG, Vaz MM, Barizon EA, Marquele-Oliveira F, Gaspar AM, Ribeiro SJ, Berretta AA (2013) Antimicrobial Brazilian propolis (EPP-AF) containing biocellulose membranes as promising biomaterial for skin wound healing. Evid Based Complement Alternat Med. 703024.

Barud HGO, Barud HS, Cavicchioli M, Amaral TS, Junior OBO, Santos DM, Petersen ALOA, Celes F, Borges VM, Oliveira CI, Oliveira PF, Furtado RA, Tavares DC, Ribeiro SJL (2015, September 5) Preparation and characterization of a bacterial cellulose/SilK fibroin sponge scaffold for tissue regeneration. Carbohydr. Polym. 128:41–51.

Brandes R, Souza LD, Vargas V, OlivEirA E, Mikowski A, Carminatti C, Al-QuresHi H, Recouvreux D (2016) Preparation and characterization of bacterial cellulose/TiO2 hydrogel nanocomposite. J. Nano Res. Sw. 43:73–80.

Buceta VR, Aguilar MR, ARriaga AMH, Blanco FG, RojAs A, Tortajada R, Jiménez RAR, Lasa BV, Prieto A (2020) Anti-staphylococcal hydrogels based on bacterial cellulose and the antimicrobial biopolyester poly (3-hydroxy-acetylthioalkanoate-co-3-hydroxyalkanoate) Int. J. Biol. Macromol. 162:1869–1879.

Cacicedo ML, Islan GA, León IE, AlVArez VA, Chourpa I, Vannier EA, Aranda NG, Riascos ZVD, Fernandez Y, Schwartz S, Abasolo I, Castro GR (2018) Bacterial cellulose hydrogel loaded with lipid nanoparticles for localized cancer treatment. Colloids Surf. B. 170:596–608.

Cai J, Zhang L (2006) Unique gelation behavior of cellulose in NaOH/urea aqueous solution. Biomacromolecules. 7(1):183–189.

Carvalho T, Guedes G, Sousa FL, Freire CSR, Santos HA (2019) Latest advances on bacterial cellulose-based materials for wound healing, delivery systems, and tissue engineering. Biotechnol J. 14(12):1900059.

Chargot MS, Cybulska J, Zdunek A (2011) Sensing the structural differences in cellulose from apple and bacterial cell wall materials by Raman and FT-IR spectroscopy. Sensors. 11(6):5543–5560.

Chen M, Kang H, Gong Y, Guo J, ZHang H, Liu R (2015) Bacterial cellulose supported gold nanoparticles with excellent catalytic properties. ACS Appl. Mater. Inter. 7:21717–21726.

Codreanu A, Balta C, Herman H, Cotoraci C, Mihali CV, Zurbau N, Zaharia C, Rapa M, Stanescu P, Radu IC, VasilE E, Lupu G, Galateanu B, Hermenean A (2020) Bacterial cellulose-modified polyhydroxyalkanoates scaffolds promotes bone formation in critical size calvarial defecTS in mice. Materials (Basel). 13(6):1433.

Dahman Y (2009) Nanostructured biomaterials and bio composites from bacterial cellulose nanofibers. J Nanosci. Nanotechnol. 9(9):5105–5122.

de Olyveira GM, dos Santos ML, Daltro PB, Basmaji P, de Cerqueira Daltro G, Guastaldi AC (2014) Bacterial cellulose/chondroitin sulfate for dental materials scaffolds. J. Biomater. Tissue Eng. 4:150–154.

Dikshit PK, Kim BS (2018) Bacterial cellulose production from biodiesel: Derived crude glycerol, magnetic functionalization, and its application as carrier for lipase immobilization. Int. J. Biol. Macromol. 153:902–911.

Duarte EB, das Chagas BS, Andrade FK, Brígida AIS, Borges MF, Muniz CR, Souza Filho MDSM, Morais JPS, Feitosa JPA, Rosa MF (2015) Production of hydroxyapatite: Bacterial cellulose nanocomposites from agroindustrial wastes. Cellulose. 22:3177–3187.

Edgar KJ, Buchanan CM, Debenham JS, Rundquist PA, Seiler BD, Shelton MC, Tindall D (2001) Advances in cellulose ester performance and application. Prog. Polym. Sci. 26(9):1605–1688.

El-Wakil NA, Hassan EA, Hassan ML, Abd El-Salam SS (2019) Bacterial cellulose/phytochemical's extracts biocomposites for potential active wound dressings. Environ Sci Pollut Res Int. 26(26):26529–26541. EMBedded silk nanofibers. J. Nanosci. Nanotechnol. 12:6139–6144.

Fontes PR, Ribeiro SJL, Gaspar AMM (2019) Bacterial cellulose/phytotherapic hydrogels as dressings for wound healing. Material Sci & Eng. 3(5):162–173.

Frone AN, Panaitescu DM, Nicolae CA, Gabor AR, TRuscA R, Casarica A, Stanescu PO, Baciu DD, Salageanu A (2020) Bacterial cellulose sponges obtained with green cross-linkers For tissue engineering. Mater Sci Eng C Mater Biol Appl. 110:110740.

Fürsatz M, Skog M, Sivlér P, Palm E, Aronsson C, SkAllberG A, Greczynski G, KHalaf H, BengTsson T, Aili D (2018) Functionalization of bacterial cellulose wound dressings with the antimicrobial peptide '-poly-L-Lysine'. Biomed. Mater. 13:25014.

Gorgieva S, Hribernik S (2019) Microstructured and degradable bacterial cellulose: Gelatin composite membranes: Mineralization aspects and biomedical relevance. Nanomaterials. 9:303.

Gupta A, Keddie DJ, Kannappan V, Gibson H, Khalil IR, Kowalczuk M, et al. (2019) Production and characterisation of bacterial cellulose hydrogels loaded with curcumin encapsulated in cyclodextrins as wound dressings. Eur. Polym. J. 118:437–450.

Gupta A, Low WL, Radecka I, Britland ST, Amin MCIM, Martin C (2016) Characterisation and in vitro antimicrobial activity of biosynthetic silver-loaded bacterial cellulose hydrogels. J Microencapsul. 33(8):725–734.

Gutierrez E, Burdiles PA, Quero F, Palma P, Moya FA, Palza H (2019) 3D printing of antimicrobial alginate/bacterial-cellulose composite hydrogels by incorporating copper nanostructures. ACS Biomater. Sci. Eng. 5(11):6290–6299.

Hamidi M, Azadi A, Rafiei P (2008) Hydrogel nanoparticles in drug delivery. Adv Drug Deliv Rev. 60(15):1638–1649.

Hawkey PM (2008) The growing burden of antimicrobial resistance. J Antimicrob Chemother. 62:1–9.

He W, Huang X, Zheng Y, Yi Sun, Xie Y, Wang Y, Yue L (2018) In situ synthesis of bacterial cellulose/copper nanoparticles composite membranes with long-term antibacterial property. Journal Biomater. Sci. Polym. Ed. 29(17):2137–2153.

Heinze T, Dicke R, KoschellA A, Kull AH, Klohr EA, Koch W (2000) Effective preparation of cellulose derivatives in a new simple cellulose solvent. Macromol. Chem. Phys. 201(6):627–631.

Hobzova R, Hrib J, Sirc J, Karpushkin E, Michalek J, JanOuskova O, Gatenholm P (2018) Embedding of bacterial cellulose nanofibers within PHEMA hydrogel matrices: Tunable stifness composites with potential for biomedical applications. J. Nanomater. 2018:1–11.

Hosseini H, Zirakjou A, Goodarzi V, Mousavi SM, Khonakdar HA, Zamanlui S (2020) Lightweight aerogels based on bacterial cellulose/silver nanoparticles/polyaniline with tuning morphology of polyaniline and application in soft tissue engineering. Int J Biol Macromol. 152:57–67.

Hu W, Chen S, Li X, Shi S, Shen W, Zhang X, Wang H (2009) In situ synthesis of silver chloride nanoparticles into bacterial cellulose membranes. Mater. Sci. Eng. C. 29(4):1216–1219.

Inoue BS, Streit S, Dos Santos Schneider AL, Meier MM (2020) Bioactive bacterial cellulose membrane with prolonged release of chlorhexidine for dental medical application. Int J Biol Macromol. 148(1):1098–1108.

Jebel FS, Almasi H (2016) Morphological, physical, antimicrobial and release properties of ZnO nanoparticles-loaded bacterial cellulose films. Carbohydr. Polym. 149:8–19.

Jia Y, Zhu W, Zheng M, Huo M, Zhong C (2018) Bacterial cellulose/hyaluronic acid composite hydrogels with improved viscoelastic properties and good thermodynamic stability. Plast. Rubber Compos. 47:165–175.

Jiji S, UdhayaKumaR S, Maharajan K, Chellan R, Chellapa M, Kadirvelu K (2020) Bacterial cellulose matrix with *in situ* impregnation of silver nanoparticles via catecholic redox chemistry for third degree burn wound healing. Carbohydr. Polym. 245:116573.

Khalid A, Khan R, Ul-IslaM M, Khan T, Wahid F (2017) Bacterial cellulose-zinc oxide nanocomposites as a novel dressing system for burn wounds. Carbohydr. Polym. 164:214–221.

Kim HJ, Jin JN, Kan E, Kim KJ, Lee SH (2017) Bacterial cellulose-chitosan composite hydrogel beads for enzyme immobilization. Biotechnol. Bioproc E. 22:89–94.

Kim J, Cai Z, Chen (2010) Biocompatible bacterial cellulose composites for biomedical application: ASME. J. Nanotechnol. Eng. Med. 1(1):011006.

Klemm D, Heublein B, Fink HP, Bohn AA (2005) Cellulose fascinating biopolymer and sustainable raw material. Angew. Chem. Int. Ed. 44:3358–3393.

Klemm D, SchUmann D, Udhardt U, Marsch S (2001) Bacterial synthesized cellulose: Artificial blood vessels for microsurgery. Prog Polym Sci. 26:1561–1603.

Klinthoopthamrong N, ChaikiaWKeaw D, Phoolcharoen W, RattanaPisit K, Kaewpungsup P, Pavasant P, Hoven P (2020) Bacterial cellulose membrane conjugated with plant-derived osteopontin: Preparation and its potential for bone tissue regeneration. Int J Biol Macromol. 149(15):51–59.

Lee SE, Park YS (2017) The role of bacterial cellulose in artificial blood vessels. Mol. Cell. Toxicol. 13:257–261.

Lemire JA, Harrison JJ, Turner RJ (2013) Antimicrobial activity of metals: Mechanisms, molecular targets and applications. Nat Rev Microbiol. 11(6):371–384.

Li Q, Yao G, Zeng X, Jing Z, Huo Z, Jin J (2012) Facile and green production of Cu from CuO using cellulose under hydrothermal conditions. Ind. Eng. Chem. Res. 51(7):3129–3136.

Liu W, Du H, Zhang M, Liu K, Liu H, Xie H, Zhang X, Si C (2020) Bacterial cellulose-based composite scaffolds for biomedical applications: A review. ACS Sustainable Chem. Eng. 8(20):7536–7562.

Liu X, Souzandeh H, Zheng Y, Xie Y, Zhong W-H, Wang C (2017) Soy protein isolate/bacterial cellulose composite membranes for high efficiency particulate air filtration. Compos. Sci. Technol. 138:124–133.

Luo LJ, Lin TY, Yao CH, Kuo PY, Matsusaki M, Harroun SG, Huang CC, Lai JY (2019) Dual-functional gelatin-capped silver nanoparticles for antibacterial and antiangiogenic treatment of bacterial keratitis. J. Colloid Interface Sci. 536:112–126.
Malheiros PS, JozAla AF, Pessoa Jr A, Vila MMDC, Balcão VM, Franco BDGM (2018) Immobilization of antimicrobial peptides from LactobaCIllus sakei subsp. sakei 2a in bacterial cellulose: StructuRAl and functional stabilization. Food Packag. 17:25–29.
Malmir S, Karbalaeia A, PourMadadib M, Hamedid J, Yazdianc F, Navaee M (2020) Antibacterial properties of a bacterial cellulose CQD-TiO2 nanocomposite. Carbohydr. Polym. 234:115835.
Mirtalebi SS, Almasi H, Khaledabad MA (2019) Physical, morphological, antimicrobial and release properties of novel MgO-bacterial cellulose nanohybrids prepared by in-situ and ex-situ methods. Int J Biol Macromol. 128:848–857.
Mohamad N, Loh EYX, Fauzi MB, Ng MH, Mohd Amin MCI (2019) In vivo evaluation of bacterial cellulose/acrylic acid wound dressing hydrogel containing keratinocytes and fibroblasts for burn wounds. Drug Deliv. Transl. Res. 9:444–452.
Mohammadkazemi F, Azin M, Ashori A (2015) Production of bacterial cellulose using different carbon sources and culture media. Carbohydr. Polym. 117(6):518–523.
Moraes PRFDS, Saska S, Barud H, Lima LRD, Martins VDCA, Plepis AMDG, Ribeiro SJL, Gaspar AMM (2016) Bacterial cellulose/collagen hydrogel for wound healing. Mater. Res. 19(1):249.
Müller A, Ni Z, Hessler N, Wesarg F, Müller FA, Kralisch D, Fischer D (2013) The biopolymer bacterial nanocellulose as drug delivery system: Investigation of drug loading and release using the model protein albumin. J Pharm Sci. 102(2):579–592. doi: 10.1002/jps.23385
Nakayama A, Kakugo A, Gong JP, Osada Y, Takai M, EraTa T, Kawano S (2004) High mechanical strength double network hydrogel with bacterial cellulose. Adv. Funct. Mater. 14(11):1124–1128.
Narh C, ChArles F, Mensah A, Qufu W (2019) Synthesis of highly stable bacterial cellulosic pocket for drug storage. Carbohydr. Polym. 206:625–632.
Noh YK, Dos Santos Da Costa A, Park YS, Du P, Kim IH, Park K (2019) Fabrication of bacterial cellulosecollagen composite sca_olds and their osteogenic effect on human mesenchymal stem cells. Carbohydr. Polym. 219:210–218.
Osorio M, Ortiz I, Gañán P, Naranjo T, Zuluaga R, van Kooten TG, Castro C (2019) Novel surface modification of three-dimensional bacterial nanocellulose with cell-derived adhesion proteins for soft tissue engineering. Mater. Sci. Eng. C. 100:697–705.
Park S, Baker JO, HimMEl ME, PArilla PA, Johnson DK (2010) Cellulose crystallinity index: Measurement techniques and their impact on interpreting cellulose performance. Biotechnol. Biofuels. 3(10):1–10.
Phutanon N, Motina K, Chang YH, Ummartyotin S (2019) Development of CuO particles onto bacterial cellulose sheets by forced hydrolysis: A synergistic approach for generating sheets with photocatalytic and antibiofouling properties. Int J Biol Macromol. 136:1142–1152.
Pirich CL, de Freitas RA, Torresi RM, Picheth GF, Sierakowski MR (2017) Piezoelectric immunochip coated with thin films of bacterial cellulose nanocrystals for dengue detection. Biosens Bioelectron. 92:47–53.
Pourreza N, Golmohammadi H, Naghdi T, Yousefi H (2015) Green in-situ synthesized silver nanoparticles embedded in bacterial cellulose nanopaper as a bio nanocomposite plasmonic sensor bios. Bioelectron. 74:353–359.
Qiu Y, Qiu L, Cui J, Wei Q (2016) Bacterial cellulose and bacterial cellulose-vaccarin membranes for wound healing. Mater. Sci. Eng. C. 59:303–309.
Romero LVC, Valls C, Valenzuela SV, Roncero MB, Pastor FIJ, Diaz P, Martínez J (2020) Bacterial cellulose: Chitosan paper with antimicrobial and antioxidant activities. Biomarcomolecules. 21(4):1568–1577.
Salari M, Khiabani MS, Mokarram RR. Ghanbarzadeh B, Kafil HS (2018a) Development and evaluation of chitosan based active nanocomposite films containing bacterial cellulose nanocrystals and silver nanoparticles. Food Hydrocoll. 84:414–423.
Saska S, Teixeira LN, Tambasco de Oliveira P, Minarelli Gaspar AM, Lima Ribeiro SJ, Messaddeq Y, Marchetto R (2012) Bacterial cellulose-collagen nanocomposite for bone tissue engineering J. Mater. Chem. 22:22102–22112.
Shao W, Liu H, Liu X, Wang S, Zhang R (2015) Anti-bacterial performances and biocompatibility of bacterial cellulose/graphene oxide composites. RSC Adv. 5:4795–4803.
Silva NH, Rodrigues AF, Almeida IF, Costa PC, Rosado C, Neto CP, Silvestre AJ, Freire CS (2014) Bacterial cellulose membranes as transdermal delivery systems for diclofenac: In vitro dissolution and permeation studies. Carbohydr Polym. 106(15):264–269. doi: 10.1016/j.carbpol.2014.02.014.

Torgbo S, Sukyai P (2019) Fabrication of microporous bacterial cellulose embedded with magnetite and hydroxyapatite nanocomposite scaffold for bone tissue engineering. Mater. Chem. Phys. 237:121868.

Treesuppharat W, Rojanapanthu P, Siangsanoh C, Manuspiya H, Ummartyotin S (2017) Synthesis and characterization of bacterial cellulose and gelatin-based hydrogel composites for drug-delivery systems. Biotechnol. Rep. 15:84–91.

Vasconcelos NF, Andrade FK, Vieira LDAP, Vieira RS, Vaz JM, Chevallier P, Mantovani D, Borges MDF, Rosa MDF (2020) Oxidized bacterial cellulose membrane as support for enzyme immobilization: Properties and morphological features. Cellulose. 27:3055–3083.

Wahid F, Duan YX, Hu XH, Chu LQ, Jia SR, Cui JD, Zhong C (2019) A facile construction of bacterial cellulose/ZnO nanocomposite films and their photocatalytic and antibacterial properties. Int J Biol Macromol. 132:692–700.

Wang J, Gao C, Zhang Y, Wan Y (2010) Preparation and in vitro characterization of BC/PVA hydrogel composite for its potential use as artificial cornea biomaterial. Mater. Sci. Eng. C. 30:214–218.

Wanna D, Alam C, Toivola DM, Alam P (2013) Bacterial cellulose: Kaolin nanocomposites for application as biomedical wound healing materials. Adv. Nat. Sci: Nanosci. Nanotechnol. 4:45002.

Wei J, Wang B, L Z, Wu Z, Zhang M, SheNg N, Liang Q, Wang H, Chen SA (2020) 3D-printable TEMPO-oxidized bacterial cellulose/alginate hydrogel with enhanced stability via nanoclay incorporation. Carbohydr. Polym. 238:116207–116217.

Wichai S, ChuySinuan P, Chaiarwut S, Ekabutr P, Supaphol P (2019) Development of bacterial cellulose/alginate/chitosan composites incorporating copper (II) sulfate as an antibacterial wound dressing. J Drug Deliv Sci Technol. 51:662–671.

Wu H, Williams GR, Wu J, Wu J, Niu S, Li H, Wang H, Zhu L' (2018) Regenerated chitin fibers reinforced with bacterial cellulose nanocrystals as suture biomaterials. Carbohydr. Polym. 180:304–313.

Wu J, Zhenga Y, Song W, Luan J, Wen X, Wu Z, Chen X, Wang Q, Guo S (2014) In situ synthesis of silver-nanoparticles/bacterial cellulose composites for slow-released antimicrobial wound dressing. Carbohydr. Polym. 102:762–771.

Xie F, Cai W, Liu Y, Li Y, Du B, Feng L, Qiu L (2015) Vaccarin attenuates the human EA.hy926 endothelial cell oxidative stress injury through inhibition of notch signaling. Int. J. Mol. Med. 35:135–142.

Yang G, Yao Y, Wang C (2017) Green synthesis of silver nanoparticles impregnated bacterial cellulose-alginate composite film with improved properties. Mater. Lett. 209:11–14.

Yuan N, Xu L, ZHang L, Ye H, Zhao J, Liu Z, Rong J (2016) Superior hybrid hydrogels of polyacrylamide enhanced by bacterial cellulose nanofiber clusters. Mater. Sci. Eng. C. 67:221–230.

Zaborowska M, Bodin A, BäckdaHl H, Popp J, Goldstein A, Gatenholm P (2010) Microporous bacterial cellulose as a potential scaffold for bone regeneration. Acta Biomater. 6(7):2540–2547.

Zhang H, Wu J, Zhang J, He J (2005) 1-allyl-3-methylimidazolium chloride room temperature ionic liquid: A new and powerful nonderivatizing solvent for cellulose. Macromolecules. 38(20):8272–8277.

Zhang W, Wang X-C, Li X-Y, Zhang L-L, Jiang F (2020) A 3D porous microsphere with multistage structure and component based on bacterial cellulose and collagen for bone tissue engineering. Carbohydr. Polym. 236:116043.

Zhao H, Kwak JH, Wang Y, Franz, JA, White JM, Holladay JE (2007) Interactions between cellulose and N-methylmorpholine-N-oxide. Carbohydr. Polym. 67(1):97–103.

Zheng W, Chen S, Zhao S, Zheng Y, Wang H (2014) Zinc sulfide nanoparticles template by bacterial cellulose and their optical properties. J. Appl. Polym. Sci. 131(19).

Zywicka A, Banach A, Junka AF, DRozd R, Fijałkowski K (2018) Bacterial cellulose as a support for yeast immobilization: Correlation between carrier properties and process efficiency. J Biotechnol. 291:1–6.

Index

F

G

H

I

K

L

M

N

O

P

R

S

T

U

V

W

X

Y